VOLUME ONE HUNDRED AND TEN

ADVANCES IN
COMPUTERS
Dark Silicon and Future On-chip Systems

VOLUME ONE HUNDRED AND TEN

Aᴅᴠᴀɴᴄᴇѕ ɪɴ
COMPUTERS
Dark Silicon and Future On-chip
Systems

Edited by

ALI R. HURSON
*Missouri University of Science and Technology,
Rolla, MO, United States*

HAMID SARBAZI-AZAD
*Sharif University of Technology,
Institute for Research in Fundamental Sciences (IPM),
Tehran, Iran*

ACADEMIC PRESS
An imprint of Elsevier

Academic Press is an imprint of Elsevier
50 Hampshire Street, 5th Floor, Cambridge, MA 02139, United States
525 B Street, Suite 1650, San Diego, CA 92101, United States
The Boulevard, Langford Lane, Kidlington, Oxford OX5 1GB, United Kingdom
125 London Wall, London, EC2Y 5AS, United Kingdom

First edition 2018

Notices
Knowledge and best practice in this field are constantly changing. As new research and experience broaden our understanding, changes in research methods, professional practices, or medical treatment may become necessary.

Practitioners and researchers must always rely on their own experience and knowledge in evaluating and using any information, methods, compounds, or experiments described herein. In using such information or methods they should be mindful of their own safety and the safety of others, including parties for whom they have a professional responsibility.

To the fullest extent of the law, neither the Publisher nor the authors, contributors, or editors, assume any liability for any injury and/or damage to persons or property as a matter of products liability, negligence or otherwise, or from any use or operation of any methods, products, instructions, or ideas contained in the material herein.

ISBN: 978-0-12-815358-1
ISSN: 0065-2458

For information on all Academic Press publications
visit our website at https://www.elsevier.com/books-and-journals

Working together
to grow libraries in
developing countries

www.elsevier.com • www.bookaid.org

Publisher: Zoe Kruze
Acquisition Editor: Zoe Kruze
Editorial Project Manager: Shellie Bryant
Production Project Manager: James Selvam
Cover Designer: Miles Hitchen

Typeset by SPi Global, India

CONTENTS

7. Introduction to Emerging SRAM-Based FPGA Architectures in Dark Silicon Era   259

Zeinab Seifoori, Zahra Ebrahimi, Behnam Khaleghi, and Hossein Asadi

PREFACE

Traditionally, *Advances in Computers*, the oldest series to chronicle of the rapid evolution of computing, annually publishes four volumes, each one typically comprised of four to eight chapters, describing new developments in the theory and applications of computing. The 110th volume entitled "Dark Silicon and Future On-Chip Systems" is a thematic volume inspired by the advances in technology and rapid increase in silicon density. However, the exponential growth in transistor count and supply voltages do not scale proportionately and hence the power consumption of the additional transistors can no longer be compensated through various circuit-level techniques. As a result, a good portion of the chips in the future will be kept in dark since we cannot afford to power up. This volume is a collection of seven chapters that are intended to address various issues within the scope of the so-called Dark Silicon.

Chapter 1, "Dark Silicon and the History of Computing," by Lotfi-Kamran and Sarbazi-Azad is an introductory chapter to this volume. It is intended to review the historical changes in the design of microprocessors, performance improvement of processors due to the increase in chip density and clock frequency, and highlighting the unavoidable emergence of dark silicon. It also shows the performance improvement of processors in the face of dark silicon.

In Chapter 2, "Revisiting Processor Allocation and Application Mapping in Future CMPs in Dark Silicon Era," Hoveida et al. articulate the importance of Network-on-Chip (NoC) design and management in offering high-performance computation in a dark-silicon-based chip multiprocessor system. In addition, they propose a scalable NoC architecture and its required management policies. By employing the proposed NoC architecture, they review the processor allocation strategy and application-to-core mapping algorithms in order to make maximum use of the NoC bandwidth and capability while meeting the power and performance goals of the hardware platform and application, respectively. Finally, the proposed algorithms have been simulated to demonstrate the system performance improvement.

Design and application of high-density chips require balancing several parameters such as performance, power consumption, and cost. This is the main theme of Chapter 3, "Multiobjectivism in Dark Silicon Age," by Rezaei et al. First the state of the art in the related area is overviewed and then a Network-on-chip-based Many-Core (MCSoC) architecture, namely shift sprinting, is introduced. In addition, an application mapping approach, called round rotary mapping, for Hybrid Wireless Network-on-Chip-based MCSoC is proposed. This approach is intended to balance the usage of wireless links by avoiding congestion over wireless routers and spreading temperature across the entire chip.

In Chapter 4 entitled "Dark Silicon Aware Resource Management for Many-Core Systems," Khdr et al. discuss about a resource management technique in the face of "Dark Silicon." The proposed scheme is intended to maximize the overall system performance under a temperature constraint. Taking the instruction and thread-level parallelisms into consideration, the scheme determines the number of active cores that should be allocated to each application and the voltage and frequency levels of these cores. The proposed technique also selects the positioning of dark cores in an attempt to facilitate dissipation of the generated heat by the active cores. Finally, the effectiveness of the proposed technique is evaluated and compared against some techniques as advanced in the literature.

Chapter 5, "Dynamic Power Management for Dark Silicon Multicore Processors," by Garg is a survey of scalable run-time dynamic power management techniques and algorithms for a homogeneous/heterogeneous multi-core platform. In addition, it presents an in-depth study of three specific power management techniques, namely: Thread Progress Equalization (TPEq), DoPpler Shift, and threshold policies.

In Chapter 6 entitled, "Topology Specialization for Networks-on-Chip in the Dark Silicon Era," Modarressi and Sarbazi-Azad propose a methodology for a Network-on-Chip platform that reduces both power consumption and communication latency. This is achieved by leveraging dark routers of partially active cores to customize the topology for active cores.

Finally, Chapter 7, "Introduction to Emerging SRAM-Based FPGA Architectures in Dark Silicon Era," by Ebrahimi et al. is a survey on evolution of SRAM-based FPGA architectures in the face of dark silicon.

We hope that readers find these articles of interest, and useful for teaching, research, and other professional activities. We welcome feedback on the volume, as well as suggestions for topics of future volumes.

ALI R. HURSON
Missouri University of Science and Technology,
Rolla, MO, United States

HAMID SARBAZI-AZAD
Sharif University of Technology,
Institute for Research in Fundamental Sciences (IPM),
Tehran, Iran

CHAPTER ONE

Dark Silicon and the History of Computing

Pejman Lotfi-Kamran*, Hamid Sarbazi-Azad*,†
*School of Computer Science, Institute for Research in Fundamental Sciences (IPM), Tehran, Iran
†Sharif University of Technology, Tehran, Iran

Contents

Abstract

For many years, computer designers benefitted from Moore's law and Dennard scaling to significantly improve the speed of single-core processors. The failure of Dennard scaling pushed the computer industry toward homogenous multicore processors for the performance improvement to continue without significant increase in power consumption. Unfortunately, even homogeneous multicore processors cannot offer the level of energy efficiency required to operate all the cores at the same time in today's and especially tomorrow's technologies. As a result of lack of energy efficiency, not all the cores in a multicore processor can be functional at the same time. This phenomenon is referred to as *dark silicon*. In this chapter, we go over the history of computing and review some of the major changes in microprocessors. Specifically, we articulate why dark silicon is inevitable and how the performance of processors can significantly be improved in the age of dark silicon.

Advances in Computers, Volume 110
ISSN 0065-2458
https://doi.org/10.1016/bs.adcom.2018.03.001

1

1. INTRODUCTION AND BACKGROUND

We are living in an era in which Information Technology (IT) is shaping our society. More than any time in history, our society is dependent on IT for its day-to-day activities. Education, media, science, social networking, etc. are all affected by IT. The steady growth in processor performance is one of the driving forces behind the success and widespread adoption of IT.

Historically, the improvement in processor performance was driven by two phenomena: Moore's law [1] and Dennard scaling [2]. Technology scaling, which refers to the technology of shrinking transistor dimensions, provided processor designers with twice transistor density every 2 years. We refer to this phenomenon as Moore's law. Moreover, the reduction in the supply voltage enabled processor designers to operate twice the number of transistors that technology offers without an increase in power consumption. This phenomenon is referred to as Dennard scaling. Taking advantage of Moore's law and Dennard scaling, computer architects improved the processing power by constantly increasing the complexity of the processor pipeline and the frequency of the processor. Decades of technology scaling allowed powerful processors with deep and aggressive Out-of-Order (OoO) pipelines and high clock frequency to emerge.

Unfortunately, improving the performance of processors with the historical approach is no longer viable. As physical restrictions slow down the reduction of the supply voltage, Dennard scaling has effectively stopped [3]. While Moore's law is still valid and the number of transistors increases by a factor of two every 2 years, the failure of Dennard scaling makes power and energy the primary constraints of processors. As such, it is no longer desirable to increase the clock frequency of processors or increase the complexity of the processor pipeline to improve performance, as these approaches are not energy efficient [4]. For this reason, we did not see a noticeable increase in the clock frequency of processors for over a decade.

As the historical approach for improving processor performance no longer works, since 2004, vendors have started to produce multicore processors using relatively aggressive OoO cores [5]. This paradigm shift was motivated by the fact that many workloads have inherent thread-level parallelism and can benefit from multiple cores. With improvement in process technology, processor vendors keep the complexity of the cores constant and use the extra transistors to increase the number of cores and the size of the last-level cache (LLC). As caches consume less energy as compared to the cores, processors use almost half of their transistor budget for the LLC with the hope that a larger cache captures a larger fraction of the data working sets and results in faster execution.

The expectation of many researchers in the field of computer architecture was that the multicore era in which the number of cores doubles every 2 years continues for several decades, and we will witness processors with up to 1000 cores. Unfortunately, multicore processors have faced several major challenges that limit their effectiveness. Some of these challenges are general and apply to all types of processors (and even all digital systems), while others are unique to multicore processors. Three of the major challenges are energy and power, off-chip bandwidth, and degree of parallelism in parallel applications. Due to these challenges, a 100-core processor may behave like a processor with just 20 cores. This means that while 100 cores are in the processor, the processor can only use 20 of them. This phenomenon is referred to as utilization wall. Due to the utilization wall, a processor cannot effectively use all of its transistors. Part of a chip that cannot be used effectively by a processor is referred to as dark silicon.

In this chapter, we review the historical ways in which processor performance has been improved. We review the challenges that caused a style of performance improvement to be replaced with another one. Specifically, we explain why the industry shifted toward having multiple cores in processors since 2004 and why we do not expect to see processors with 1000 cores in the current CMOS technology. We review in detail each of the three major challenges of multi- and many-core processors and explain how it results in chips with dark silicon. We also briefly talk about various ways in which each of these challenges may get addressed.

2. THE SINGLE-CORE ERA

Since the invention of microprocessors in 1971 till around 2004, we are in the single-core era. In this period of time, we witnessed a significant improvement in processor performance. The improvement in processors' performance mostly comes from two sources: (1) technology advancement and (2) architectural and microarchitectural advancements. Technology advancement refers to two related phenomena, technology scaling and Dennard scaling. Technology scaling refers to the process of shrinking transistor dimension that results in (1) more transistors in a fixed area and (2) faster transistors. Dennard scaling refers to the process of reducing the supply voltage of more advanced technologies in order to operate more transistors with the same power as the older technology, as we explain in more detail later.

As a result of significant improvements in technology, we moved from the Intel 4004 microprocessor in 1971 [6], which only had 2300 transistors, to the Intel Pentium 4 in 2004 [7], which had 125M transistors. This essentially means that the number of transistors doubled every 2 years since 1971 and Moore's law was valid in the whole duration between 1971 and 2004. Moreover, while the frequency of Intel's 4004 microprocessor was only 740 kHz, the frequency of Intel Pentium 4 was 2.4 GHz. This means that frequency of processors has increased by a factor of 3240 in just 33 years. The massive increase in frequency is a result of having faster transistors due to technology scaling and architectural and microarchitectural advancements. Advanced processors use deep pipelining technology to significantly reduce the delay of logic between to memory elements. Consequently, such processors can be clocked with high frequency signals.

The improvement in clock frequency is not the only factor contributing to the performance improvement of processors in 1971–2004. The extra transistors have been used by computer architects to build more functional units, better branch predictors, larger caches, advanced out-of-order execution units, etc. to further improve the performance. All these factors contributed to $100,000 \times$ performance improvement of processors.

Several related factors contributed to the end of the single-core era. The performance improvement of a single-core processor comes mostly from (1) increasing the frequency, (2) larger and hence more effective caches, and (3) instruction-level parallelism (ILP). The goal of ILP is to find independent instructions in the sequence of instructions and execute them in parallel. Around 2004, high-performance processors could execute up to four instructions in parallel. As the average size of a basic block (i.e., the sequence of instructions between two branch instructions) is six instructions [8], there is unlikely that enabling a processor to execute more than four independent instructions in a cycle leads to significantly higher performance. To make things worse, not only making a processor more aggressive to execute more instructions in a cycle leads to diminishing return due to lack of enough ILP, but also a more aggressive processor consumes significantly more power.

The main reason behind the end of single-core era is power consumption. The main equation that defines the power consumption of a processor is shown in Eq. (1) [9].

$$P = ACV^2f + \tau AVI_{\text{short}} + VI_{\text{leak}} \tag{1}$$

There are three components in this equation. The first one measures the dynamic power consumption, the second one measures the transient power, and the last one measures the leakage power. The dynamic power, which is the power consumes to charge or discharge the capacitance of the gate, is linearly proportional to the frequency (f), the gate capacitance (C), and the activity factor (A). Moreover, it is quadratically proportional to the supply voltage (V). The transient power, which is the power consumes when the output switches, is proportional to the duration of the transition (τ), activity factor (A), supply voltage (V), and the short current between supply voltage and ground (I_{short}). Finally, the leakage power, which is the power consumes due to leakage current, is proportional to the supply voltage (V) and the leakage current (I_{leak}).

In the CMOS technology, the transient power is negligible, and up until 2004, the leakage power was also negligible. As such, the main factor contributing to the power consumption in the single-core era is the dynamic power, which is described by the first component of Eq. (1). In the single-core era, as technology advanced and more transistors became available, it was possible to tune the parameters of Eq. (1) to operate twice as many transistors that the newer technology offered with the same power of the older technology. In a newer technology, the capacitive load of gates (C) is smaller than the older technology. The smaller capacitive load can compensate the larger frequency at which the transistors in the more advanced technology operate. In order to compensate for having twice as many transistors as the older technology, circuit designers reduced the supply voltage (V) to decrease the power consumption. As dynamic power is quadratically proportional to the supply voltage, a slight decrease in it could result in significant power saving.

Unfortunately, the maximum frequency at which the transistors can operate becomes lower when we decrease the supply voltage. Eq. (2) describes the relationship between the maximum frequency and the supply voltage [9].

$$f_{\text{max}} \propto (V - V_{\text{th}})^2 / V \tag{2}$$

In Eq. (2), V_{th} is the threshold voltage and V is the supply voltage. Eq. (2) tells us that as supply voltage decreases, the maximum frequency also decreases. As increasing the frequency is one of the major goals of processor designers in the single-core era, they often reduced the threshold voltage with the supply voltage in a new technology to increase the maximum frequency. Not only decreasing the threshold voltage along with the supply voltage in a new technology helps getting higher frequencies, but also it is necessary for the correct operation of the low-power circuits [9].

Eq. (1) indicates that reducing the supply voltage by a factor of two, decreases the power consumption by a factor of four. One the other hand, Eq. (2) indicates that maximum frequency has linear dependence to the supply voltage. Therefore, reducing the supply voltage by a factor of two reduces the maximum frequency by a factor of two. This means that if a task can be perfectly divided into two parts, one can execute the two subtasks on two processors operating at $f_{max}/2$ with supply voltage of $V/2$, instead of one processor that operates at f_{max} with the supply voltage of V, essentially with the same performance and half the power usage. This observation is used to justify the move to the multicore era.

One of the main reasons behind the end of the single-core era is captured by the following equation, which describes the relationship between the threshold voltage and the leakage current [9].

$$I_{leakage} \propto \exp\left(-V_{th}/35mV\right) \tag{3}$$

Eq. (3) essentially says that the leakage current is inversely proportional to the threshold voltage. As the threshold voltage decreases, the leakage current increases. This makes leakage power of Eq. (1) more and more important. Over the many years of the single-core era, both the supply voltage and the threshold voltage have been reduced significantly, to the points where in the modern technology, up to 40% of the total power consumption is due to leakage power [10]. As it is no longer possible to sufficiently reduce the threshold voltage, the supply voltage cannot be reduced as well. Therefore, as the technology advances and more transistors become available, we cannot operate them with the same power of the older technology, as we used to. In the end of the single-core era, power and energy are the main constraints of digital systems.

One way to get higher performance without commensurate increase in power is parallel processing. As noted earlier, if a task is parallelizable, one can run it on two or more processors in parallel with the same performance and lower power usage. As it was no longer possible, due to power constraints, to use the extra transistors with the maximum frequency, a paradigm shift happened around 2004. Instead of having a single processor that operates at maximum frequency, the designers use the extra transistors that a technology offers to increase the number of cores in a processor. Each core operates in the same frequency as the older technology. Hopefully, the reduced capacitive load and the very slight reduction in the supply voltage, as we do not need to operate in the maximum frequency that the new technology offers, let us operate twice as many transistors with the same power or just slightly more power as the older technology. And that's how the multicore era began.

3. THE MULTICORE ERA

As of 2004, most vendors started to produce multicore processors. In the end of the single-core era, we have reached the end of ILP and also power consumption started to become a major constraint. Therefore, single-core processors could not offer a reliable way going forward. The move to the multicore era was motivated by the fact that many workloads have inherent parallelism and the fact that by taking advantage of parallelism, we can essentially run a job with lower power consumption, as explained in Section 2.

In the multicore era, as technology advances and more transistors become available, the vendors use the extra transistors to put more cores into a processor. So just like good old times, processor performance is expected to double every technology generation. The only difference is that in the multicore era, it is the responsibility of the applications to take advantage of multiple cores in a processor. As a single core in a processor does not offer significantly higher performance as compared to a core in the older technology, if an application cannot benefit from the extra cores in a processor, unlike the single-core era, technology advancement does not directly contribute to higher performance. It is the responsibility of the applications to take advantage of the extra cores in a processor in order to offer higher performance. Unlike the single-core era, application developers need to redesign their applications every technology generation for the applications to be able to effectively use the extra cores that the new technology offers.

In the new era, a processor consists of several identical cores that are connected together using a network-on-chip (NoC). All the cores usually share a LLC, and a coherence directory is responsible to make sure that the cores access the latest value of a memory location in their private caches. As technology advances and more transistors become available, the extra transistors are used to place more cores in a processor. The processors in the multicore era are homogeneous in a sense that all the cores in a processor are identical.

In the beginning of the multicore era, the expectation of many in the computer industry was that the multicore era will last for several decades, just like the single-core era, and we will see processors with 1000s of cores. Consequently, many argued for changing the curriculum of computer science and engineering in order to teach all the courses with the perspective of parallelism and parallel processors to prepare scientists and engineers for the new era. Moreover, people in the research community started to take a look at the components in a multicore processor that will become the bottleneck should we have 1000 cores in processors. Several proposals have been offered for components like NoC or directory that can scale to 1000 cores.

Unlike the initial enthusiasm for the multicore era, several major constraints directed homogenous multicore processors to the utilization wall [11,12]. Utilization wall means that applications cannot effectively use all the cores in a processor. Due to the utilization wall, as we discuss in more detail later in this section, we cannot effectively use all the silicon budget available to a processor and hence we moved to the dark silicon era. The main reasons for the utilization wall are (1) lack of parallelism, (2) off-chip bandwidth constraint, and (3) power and energy constraint. We discuss each of these reasons in more details in the rest of this section.

3.1 Lack of Parallelism

Moving from single-core processors to multicore processors was mainly motivated by the fact that many applications are inherently parallel and can benefit from execution on multiple cores. While this intuition is generally true, different applications offer different levels of parallelism. The fraction of an application that can be executed in parallel is referred to as its level of parallelism. This model is very simple and assumes that parallel fraction of an application is infinitely parallelizable. With this simple model, the level of parallelism of different applications cover a full range between 0% (not parallelizable) and 100% (fully parallelizable).

It is clear that multicore era is not useful for applications with limited parallelism. However, even for applications that are highly parallelizable, the serial fraction of applications (i.e., the fraction that is not parallelizable and must be run serially on a single core) ultimately limits the performance that one can get using parallel execution. A simple formula captures the

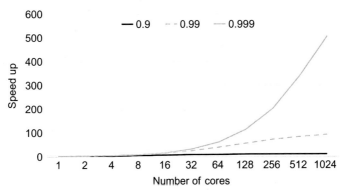

Fig. 1 Speedup of three applications with 0.9, 0.99, and 0.999 parallel fraction on various numbers of cores.

relationship between the level of parallelism, core count, and speedup. This formula is often referred to as Amdahl's law [13].

$$S = 1/((1 - p) + (p/n)) \tag{4}$$

Amdahl's law assumes that an application can be divided into two parts: serial part and parallel part. In Eq. (4), p is the fraction of the execution time of parallel part to the whole execution time when the application runs on a single core, n is the number of cores, and S is the speedup that one gets if the application runs on n cores.

Fig. 1 shows the speedup for various numbers of cores for three applications with level of parallelism of 0.9, 0.99, and 0.999. The figure shows that the line corresponding to the application with 90% parallelism is essentially flat. The speedup of this application with 1024 cores is just 10×. For an application with 99% parallelism, we observe higher speedup as compared to the application with 90% parallelism. However, even for such an application, the speedup is just 91× when the application executes on 1024 cores. The highest speedup is observed for an embarrassingly parallel application with 99.9% parallelism. The speedup for this application is significantly higher than the other two applications. Even for such an embarrassingly parallel application, the speedup is just 506× when it runs on 1024 cores. This means that most of the cores (i.e., 1024–506 = 518) are not utilized when such an application runs on 1024 cores.

Unfortunately, usually achieving a level of parallelism of 99% is difficult. Many parallel applications offer less than 90% parallelism. Making such applications more parallel is extremely difficult and requires enormous effort. Moreover, Fig. 1 shows that it is becoming more and more difficult to utilize all the cores of a multicore processor as the number of cores increases, even for embarrassingly parallel applications. While there are applications with 100% parallelism (e.g., some data center cloud applications like Web Search and media streaming), many parallel applications offer parallelism below 100%. Fig. 1 shows that such applications are inherently incapable of utilizing all the cores of a many-core processor.

When there are only few cores available, the inherent parallelism in applications is sufficient to utilize the cores. However, when the number of cores increases, even many embarrassingly parallel applications are incapable to utilize all the cores. This phenomenon leads to the utilization wall in which applications cannot utilize all the cores effectively. In order to increase the utilization, we need to rewrite the applications to offer higher levels of parallelism. Unfortunately, this task is difficult and time-consuming, and even sometimes impossible.

3.2 Off-Chip Bandwidth Constraint

The second major factor behind underutilization of cores in a many-core processor is the off-chip bandwidth. While initially multicore processors had only two cores, today's processors can easily fit 16 cores. As Moore's law is still valid and the number of transistors in unit area is expected to double every 2 years, the number of cores on chip can be increased even further. The larger the number of cores on chip, the higher the demand on the off-chip memory.

The datasets of applications are stored in the main memory. Applications need to go to the main memory to access the data. As main memory is located off the chip, data accesses need to go outside the chip to reach the memory. Unfortunately, pin count does not grow proportional with the number of cores [14]. While the number of transistors and hence the number of cores is expected to double every 2 years, the rate of increase in pin count is much smaller. For cores in a multicore processor to do useful work, they need to access data in the main memory and hence need off-chip bandwidth. As the pin count is growing much slower than the core count, the off-chip bandwidth dedicated to a core is decreasing every technology generation. This means that at some point, there is not enough off-chip bandwidth to make all the cores busy. Consequently, some of the cores in a multicore processor need to be kept idle for the rest of the cores to have enough off-chip bandwidth.

Off-chip bandwidth being a limited resource, not all the cores can be utilized at the same time. Hence in the future, several cores of a multicore processor may remain idle due to bandwidth limitation should the multicore era continues. Fortunately, there are techniques to delay or completely mitigate the bandwidth wall. We discuss some of these techniques later in this chapter.

3.3 Power and Energy Constraints

The third and the most fundamental constraint that limits multicore processors is power and energy. As we discussed previously, a parallel application can be executed on multiple processors with the same performance and half the power usage of a single processor by lowering both the frequency and the supply voltage. However, the application must be 100% parallel. Moreover, the technique of lowering frequency and the supply voltage cannot be used indefinitely. For proper operation of circuits, the supply voltage must be three times as large as the threshold voltage [15]. The threshold voltage cannot be reduced indefinitely, because it exponentially increases the leakage current and power [9].

As most applications are not fully parallel and the supply voltage cannot be reduced indefinitely, power and energy are major constraints of processors even in the multicore era. As the number of cores in multicore processors increases, the power and energy usage of the processors also increases. This means that due to power constraints, not all the cores in a processor can be functional at the same time. Using PARSEC benchmarks, Esmaeilzadeh et al. [11] showed that 21% of a chip must be turned off at 22 nm technology node. This number goes up to 50% at 8 nm. Not only power is a limiting factor for traditional parallel applications like PARSEC, Hardavellas et al. [12] showed that even for server applications, which are embarrassingly parallel, a large portion of the chip silicon have to be powered down.

4. THE DARK SILICON ERA

Due to the three major constraints that we discussed in the previous section, not all the cores in a many-core processor may work at the same time. We refer to this phenomenon as the utilization wall. Because of the utilization wall, some parts of a chip must remain idle while the rest of the chip is performing the computation. In the literature, parts of a chip that are idle due to the utilization wall are referred to as dark silicon. As the utilization wall is prevalent today, we are now in the dark silicon era.

In the dark silicon era, not all cores of a processor can be active at the same time. This limitation essentially calls for a heterogeneous architecture for a multicore processor in the dark silicon era. It does not make sense to increase the number of identical cores in a processor when we cannot use them all at the same time. The dark silicon era calls for heterogeneous architectures in which every component is specialized to perform a set of tasks very efficiently.

Note that even in homogenous multicore processors, due to power limitation, designers usually did not operate a processor with the maximum frequency that is allowed in a technology. While certainly such limitations are an early indication of dark silicon in processors, we prefer to categorize homogenous multicore processors as part of the multicore era. In this perspective, heterogeneous multicore processors are the architecture of choice for the dark silicon era. In the rest of this section, we discuss various solutions that have been proposed for the three main constraints of the previous section.

4.1 Solutions for Parallelism

Multicore processors need parallel applications that can use them efficiently. As the number of cores in a multicore processor is expected to increase every few years, applications also need to change to benefit from the extra cores available in the processor. This is essentially in contrast with the single-core era in which the same programs could execute much faster on more advanced processors.

As the number of cores increases, applications need to significantly reduce the serial fraction so that they can benefit from the extra cores, as is evident from Fig. 1. For this reason, many experts called for changes to the undergraduate curriculum to solve the parallel programming challenges for multicore processors. Unfortunately, parallel programming is even more difficult and error prone than standard programing. This means that this task is very difficult and even excellent parallel applications eventually hit the parallelism wall for processors with many cores, as shown in Fig. 1 (except for applications that are 100% parallelizable like Web search or media streaming).

In addition to the inherent parallelism wall, as the number of cores in a processor increases, some components of the processor become the bottleneck for execution, even for applications with 100% parallel fraction. These components are mostly shared among all the cores in a processor and include LLC, coherence directory, and NoC. In the rest of this section, we go over each component and review some of the techniques for making the components scalable.

4.1.1 Last Level Cache

LLC is one of the components that heavily influences the performance of workloads running on multicore processors. Many piece of prior work proposed techniques to reduce the delay of the LLC. Kim et al. were the first to notice that nonuniform cache architecture offers lower average delay as compared to a uniform cache when the delay is dominated by wire delay [16]. Later, Huh et al. showed that a nonuniform cache is suitable for multicore processors [17]. In a nonuniform cache architecture, unlike a uniform cache, the cache is splitted into several slices. Each slice is located next to a core (or group of cores) in a multicore processor. From the perspective of a given core, some slices are closer and other slices are further away. The delay of accessing the cache not only depends on the delay of accessing a cache slice but also the delay of reaching the cache slice. Therefore, depending on which cache slice needs to be accessed, the delay of the cache is different (nonuniform). On average, a large nonuniform cache offers a lower delay as compared to a large uniform cache.

In a nonuniform cache architecture, the cache slices can be either private or shared. In a private setting, each core in a multicore processor only has access to the LLC slice that is next to it (or the slice is private to the core). In a private cache, the access to the cache is fast because a core only attempts to access the nearby slice. The downside of a private cache is that a core only can access a portion of the cache capacity. In a shared setting, however, a core can access all cache slices. The delay of a cache access is higher than a private cache because some of the accesses are to the further away cache slices but all the cores can benefit from the whole cache capacity.

There have been many pieces of work that advocated for a hybrid (adaptive) organization to offer the benefits of both private and shared caches [18–25]. Beckmann evaluated nonuniform cache architectures [25] and concluded that dynamic migration of blocks increases performance but requires smart lookup algorithms and may cause contention. Kandemir et al. proposed migration algorithms for the placement of each cache block in LLC slices [26], and Ricci et al. proposed smart lookup mechanisms using Bloom filters [27]. Chang et al. proposed a private cache organization that steals capacity from neighboring private slices [19]. Zhang advocated for a shared substrate that creates local replicas to reduce access latency [21]. Beckmann proposed an adaptive design that dynamically adjusts the probability by which read-only shared blocks are allocated at the local slice [18].

Reactive NUCA (R-NUCA) [24] is a cache block placement for nonuniform cache architectures. R-NUCA classifies cache blocks into three categories: private data, shared data, and instructions. In order to reduce the classification overhead, R-NUCA classifies cache accesses at the page level. R-NUCA places private cache blocks into the nearest cache slices, shared cache blocks into all cache slices, and instruction blocks into four nearby slices. With R-NUCA, shared data blocks benefit from the whole cache capacity, and private data and instruction blocks benefit from low access latency, as they are mapped to the nearby cache slices. It has been shown that R-NUCA offers performance within 5% of an ideal cache in a 16-core processor.

Many pieces of prior work showed that size of the LLC should be minimal for many important classes of applications [28,29]. Kgil et al. [30] proposed eliminating LLCs and devoting their area to cores, while compensating for the increase in memory bandwidth pressure through 3D-stacked DRAM. As many applications have significant LLC resident instruction working sets, Lotfi-Kamran et al. [31] suggested using a small LLC to capture the instruction working set while using the rest of the silicon real estate to increase the number of cores.

4.1.2 Coherence Directory

Coherence directory is one of the components in a multicore processor that is shared among all the cores and may become the bottleneck for processor performance as the number of cores increases [32]. As every core in a multicore processor has a private cache (usually L1 caches) and, as such, a piece of data may get updated in several private caches, coherence directory is responsible for delivering the latest copy of a piece of data, should a core request for it. A coherence directory usually keeps track of which private cache has a particular piece of data. There have been many proposals to address several coherence directories' shortcomings [32–49]. In what follows, we review some of the most influential pieces of prior work.

Traditional directory mechanisms are not scalable. Duplicate-tag directories [33,34] maintain a copy of the tags in private caches. While duplicate-tag directories have low storage overhead, their power usage becomes prohibitive due to its high associativity. Sparse directories [35], on the other hand, require low associativity but their storage overhead is prohibitive. The storage overhead in Sparse directories is due to (1) overprovisioning the number of entries to reduce conflicts and lower the performance degradation and (2) bits necessary to encode the sharer-set of a given piece of data.

To lower the conflict and hence reduce the overprovisioning factor, Cuckoo directory [32] benefits from a low associative (3- or 4-way) structure whose address bits go through different hash functions, one for each way. Unlike the set-associative organization that always picks a replacement victim from a small set of candidates, the Cuckoo directory displaces victims to alternate nonconflicting ways, practically never resorting to eviction. It has been shown that Cuckoo directory significantly reduces the storage overhead of Sparse directories caused by overprovisioning the number of entries. To reduce the storage requirements due to encoding the sharers, SCD [36] uses a variable number of directory tags to encode sharers: cache blocks with one or few sharers use a single tag, and widely shared cache blocks use additional tags.

4.1.3 Network-On-Chip

One of the shared components in a multicore processor that heavily influences the performance is NoC. A NoC consists of many shared busses between cores and several routers that orchestrate the passage of packets originating from different cores on the shared busses so that (1) every bus is granted to only one packet at any point in time and (2) packets ultimately reach the destination. The need for low-latency NoC is wildly known in the community. There have been many proposals to reduce the delay of the NoC. In the rest of this section, we briefly review some of the main ideas.

The proposals for reducing NoC delay target reducing (1) hop count, (2) blocking delay, and (3) per-hop latency. Reducing the hop count is one of the major goals of many low-latency NoC proposals. Many pieces of prior work achieved this goal by proposing a topology for the NoC that offers low diameter [50–52]. Others considered reconfigurable networks [53], and even standard mesh topology that is equipped with extra irregular links [54,55] to lower the hop count.

One of the major techniques to reduce the packet blocking delay is adaptive routing. An adaptive router attempts to adaptively route packets over less congested paths. Some adaptive routing algorithms only consider local congestion metrics [56,57], while others consider both local and global metrics [58–63]. While global adaptive routing algorithms are more complicated than their local counterparts, they often make more appropriate routing decisions [58]. Moreover, arbitration also influences the packet blocking delay. As such, many proposals attempt to reduce the blocking delay by a better arbiter [64–71].

Per-hop (router) latency reduction is one of the widely used ways to decrease network latency. Decreasing router latency involves cutting down or bypassing the pipeline stages of routers. Standard routers have several stages where every stage requires one clock cycle. Some proposals advocate for a single-stage router [72] by applying extensive precomputation techniques to forward packets in a single cycle under low traffic. Moreover, bypassing a router [73–75] enables packets to go over a hop in one cycle on preestablished paths. In what follows, we review some of the major pieces of prior work.

In flit-reservation flow control [76] a control packet goes into the network a few cycles before the data packet to allocate buffers and channel bandwidth. A control packet is responsible for one or multiple flits (i.e., packet segment). Bufferless NoCs also lowers the number of stages in the router pipeline. Bufferless NoC always sends a received packet to an output port [77,78]. If the desired output port is busy, the packet will be sent to another output. Consequently, such routers do not need buffers in their input ports. Bufferless NoCs usually have support for single-cycle packet forwarding. Unfortunately, the single–cycle packet forwarding comes with packet deflection or dropping when the preferred output port is busy. Packet deflection or dropping increases the network latency under moderate traffic loads.

To lower per-hop latency, circuit-switched NoCs have also been considered in the literature [79,80]. In circuit-switched networks, packets do not need to go through buffering, routing, arbitration, and flow control once circuits are set up. Unfortunately, this switching mechanism usually lowers the performance because of long circuit setup delay and poor bandwidth utilization. Schemes based on the time–division multiplexing mitigate the low bandwidth utilization of circuit switching [80], but at the cost of increasing the complexity of the NoC. Proactive circuit-switching [79] is a proposal to hide long circuit setup time. In this proposal, a request packet reserves a circuit for its corresponding response packet as it goes toward the destination. Although preallocation reduces the delay for those packets that pass through the circuits, this mechanism requires multiple NoC planes. In addition, early reservation of circuits leads to the underutilization of network bandwidth.

In another piece of prior work, a circuit-switched memory access (CIMA) NoC [81] is proposed that preallocates circuits for long response packets during the time interval in which the LLC is preparing the data. A similar piece of prior work [63] allocates circuits in a single-cycle multihop fashion under two conditions (1) the time interval in which the LLC is preparing the data and (2) the time interval in which a packet is waiting in the network. Finally, NOC-Out is a NoC that is tuned to the characteristics of scale-out workloads. NOC-Out benefits from high-radix flattened butterfly topology to reduce hop count and low latency simple routers to reduce per-hop latency [82].

4.2 Solutions for Off-Chip Traffic

As the number of cores increases, they generate more and more off-chip traffic. The number of cores increases with a much larger rate as compared to the number of pins, and as such, the off-chip traffic becomes a bottleneck for multicore processors. This bottleneck eventually prevents multicore processors from reaching the peak performance. Fortunately, there are ways to deal with the off-chip traffic. One way to reduce the off-chip traffic is to use large LLCs in processors. While large LLCs can slightly reduce the off-chip traffic, due to the mismatch between their size and the size of the datasets in modern applications, they are incapable of mitigating the off-chip traffic [28,29].

Researchers have proposed using large die-stacked DRAM to mitigate the off-chip bandwidth problem [83–92]. Fortunately, recent progress in 3D technologies makes integration of a sizeable amount of DRAM possible. Die-stacked DRAM can provide several gigabytes of storage [93]. When 3D-stacked technology is combined with "2.5D" silicon interposer-based integration, where multiple DRAM stacks can be placed in the same package [94], the in-package DRAM capacity increases even further. Fortunately, the size of the on-chip DRAM can be significantly larger than the size of the largest LLCs that are available today, and hence can be used to mitigate the off-chip bandwidth problem.

Most proposals advocate for using the on-chip DRAM as a large DRAM cache because the size of the datasets in modern applications is significantly larger than the size of the on-chip DRAM [83]. A large DRAM cache needs to have several features to be highly efficient. First, it needs to offer low miss ratio. Second, its access latency must be minimum. Third, it should not require massive tag storage on the active die. It has been shown that spatial locality plays an important role in reducing the miss ratio of large DRAM caches [83]. Therefore, highly efficient DRAM caches are page-based caches (i.e., caches with large cache blocks) to benefit from the spatial locality in the accesses to the cache and hence minimize the miss ratio.

Unfortunately, a page-based cache significantly increases the off-chip traffic because upon a miss, it brings into the cache a page that contains many blocks, and only some of the blocks are used during the life-time of the page in the cache [83]. To reduce the off-chip traffic of page-based caches, prior work [83] takes advantage of a predictor for identifying the blocks within a page that will be used while the page is in the cache (or the footprint of the page). Upon a cache miss, only those blocks within the requested page that are predicted to be used will be brought into the cache, effectively reducing the off-chip traffic.

Large DRAM caches require large tag arrays. The tag is so large that cannot be placed on the active die and instead should be placed on DRAM [84]. The naive implementation of such a cache requires two serialized accesses to the DRAM cache to get a cache block: (1) one access to read the tag and identify the location of the piece of data (i.e., way) and (2) one access to read the piece of data. To reduce the cache access latency, recent work [85] takes advantage of a way predictor to access the requested block with two overlapped read operations for tag and data.

4.3 Solutions for Power Consumption

Power consumption is the main constraint that calls for the end of the multicore era. Unlike lack of parallelism that only affects applications that are not 100% parallelizable, power consumption affects all applications. Unlike off-chip bandwidth constraint that can be mitigated with 3D-stacked technology, there is no single technology that can mitigate the power usage constraint of multicore processors.

While there is no single technology to resolve the power wall, there are several approaches that can reduce the power consumption of processors. All these approaches, in one way or another, call for a heterogeneous multicore architecture. As power consumption does not let all the cores in a processor be active at the same time, and there is no technology that magically reduces the power consumption of processors (something like 3D stacking but for reducing the power consumption), the power wall calls for a heterogeneous architecture in which each component is extremely efficient at executing certain classes of applications or certain parts of an application. In this section, we introduce three widely recognized approaches for decreasing the power consumption, namely, integration, specialization, and approximation.

4.3.1 Integration

Traditionally, systems were composed of three discrete components: processors and caches, fast controller, and slow controller. Fast controller, which sometimes is called *northbridge*, is responsible for connecting the processor to DRAM and graphic controller. On the other hand, slow controller, which sometimes is called *southbridge*, is responsible for connecting the processor to input/output (I/O) devices. As process technology gets more mature, the three discrete components are integrated on the same chip. The integration enables lowering the power consumption, in addition to increasing the performance, because it lowers the power consumes for off-chip communication. In earlier systems, every time a processor needed to access fast or slow controller, it needs to communicate with an off-chip component, which consumes lots of power. Current processors, however, integrate fast and slow controller on chip, which requires far less power for accessing them.

As power becomes more important, more components are integrated in a system on chip (SoC). Sun Niagara and the Tilera Tile64 are examples of general-purpose processor SoCs. Niagara 2 integrates a PCIe controller, two 10 Gb Ethernet ports, and many heavily multithreaded cores [95]. Companies that use ARM cores to build server processors also integrate uncore components in the server chip. For example, Marvell Armada XP [96] has an ARM-based quad-core SoC that also integrates network interfaces and storage controllers. Cavium ThunderX [97] also includes DDR controller, SATA, PCIe, 10/40/100 Gb Ethernet. The integration helps lowering the communication power usage.

While several research proposals also advocate for integrating more components on chip to lower power usage [98–103], integration can only slightly reduce the power usage. The power wall, which is a fundamental problem for digital systems, cannot be mitigated just by integration alone. Digital systems need to use other techniques along with integration to significantly reduce the power usage.

4.3.2 Specialization

The power usage of multicore processors reached a point where it is no longer possible to operate all the cores at the same time. The high power–usage of multicore processors is partially due to high power usage of general-purpose cores. While general-purpose cores can run diverse classes of applications, such cores are not efficient at executing them. It has been shown that ASIC is 500× more energy efficient than general-purpose cores [104]. As such, there is a strong consensus in the computer architecture community that future's processor chips should include a diverse class of accelerators where each accelerator can execute certain types of functions extremely efficiently. Unlike general-purpose cores that can execute any function, an accelerator can only execute a limited class of functions. However, accelerators are extremely energy efficient. As an accelerator can only execute a limited class of functions, future chips should include many different accelerators to be able to execute a diverse class of functions.

Many pieces of prior work proposed general-purpose accelerators for energy efficiency [104–108]. Most of these proposals attempt to improve the efficiency of the memory hierarchy. Unfortunately, these techniques are less applicable to modern applications due to the deep software stacks and vast datasets. On the other hand, many proposals suggest accelerators for a limited class of applications (e.g., database, graph processing, etc.) [109–111]. These accelerators usually offer higher efficiency as compared to their general-purpose counterparts. Other work proposed techniques to accelerate a large-scale application [112] or automatically generate accelerators [113]. As graphics processing units are extremely efficient at executing certain types of applications (beyond image processing) [114,115], many argue for using them along the CPU cores for acceleration of applications [116,117]. FPGAs are also a favorite choice for implementation of reconfigurable accelerators [118].

Finally, many research proposals advocate for bringing the computation closer to data for increasing the performance and decreasing the power consumption. Traditional Processing-In-Memory systems [119–125] integrate logic into the memory die to enable lower data access latency and higher memory bandwidth as well as lower communication power usage. Unfortunately, these systems suffer from high manufacturing cost, low yield, and limited mainstream adoption since memory and logic technologies are designed for different purposes. Recently, a wealth of architectures for near data processing have been proposed using 3D/2.5D stacking technologies [111,126–139].

4.3.3 Approximation

Many widely used applications like recognition, gaming, data analytics, weather prediction, and multimedia are amenable to imprecise computation [140–143]. For such applications, some variation in output is acceptable and some degradation in the output quality is tolerable. This characteristic provides a unique opportunity to devise approximation techniques that trade small losses in the quality of results for significant gains in performance and efficiency.

Recent work has explored a variety of approximation techniques to improve performance and lower the power usage that include: (a) approximate storage designs [144,145] that trades quality of data for reduced energy [144] and longer lifetime [145], (b) voltage over scaling [146–148], (c) loop perforation [149–151], (d) loop early termination [152], (e) computation substitution [140,143,152,153], (f) memoization [141,142,154], (g) limited fault recovery [150,155–159], (h) precision scaling [160,161], (i) approximate circuit synthesis [162–168], (j) neural acceleration [169–175], and approximate benchmark suite [176].

5. CONCLUSION

Computer industry took advantage of Moore's law and Dennard scaling for several decades and significantly improved the performance of single-core processors. Around 2004, due to the failure of Dennard scaling, power and energy became the major constraint of digital systems. To offer higher performance without significant increase in power, computer industry moved toward multicore processors. While multicore processors lessen the power consumption of processors for several years, the computer industry is now in the dark silicon age where not all the cores of a processor can be functional with the highest power at the same time. In this chapter, we reviewed the main reasons behind the existence of dark silicon and how the speed of processors can be increased in the age of dark silicon.

REFERENCES

[1] G.E. Moore, Cramming more components onto integrated circuits, Electronics 38 (8) (1965) 114–117.
[2] R.H. Dennard, F.H. Gaensslen, H.-N. Yu, V. Leo Rideout, E. Bassous, A.R. Leblanc, Design of ion-implanted MOSFET's with very small physical dimensions, IEEE J. Solid State Circuits 9 (5) (1974) 256–268.
[3] R.H. Dennard, J. Cai, A. Kumar, A perspective on Today's scaling challenges and possible future directions, Solid-State Electron. 51 (4) (2007) 518–525.
[4] F.J. Pollack, in: New microarchitecture challenges in the coming generations of CMOS process technologies, Proceedings of the 32nd Annual ACM/IEEE International Symposium on Microarchitecture, 1999. (Keynote Address)(Abstract Only). Page 2.
[5] D. Geer, Chip makers turn to multicore processors, Computer 38 (5) (2005) 11–13.
[6] F. Faggin, M.E. Hoff, S. Mazor, M. Shima, The history of the 4004, IEEE Micro. 16 (6) (1996) 10–20.
[7] Intel, Intel Introduces The Pentium 4 Processor. (Online). https://web.archive.org/web/20070403032914/http://www.intel.com/pressroom/archive/releases/dp112000.htm.
[8] V. Escuder, R. Duran, R. Rico, Analysis of x86 ISA condition codes influence on superscalar execution, in: S. Aluru, M. Parashar, R. Badrinath, V.K. Prasanna (Eds.), High Performance Computing—HiPC 2007. HiPC 2007, Lecture Notes in Computer Science, vol. 4873, Springer, Berlin, Heidelberg, 2007.

[9] T. Mudge, Power: a first class design constraint for future architectures, in: M. Valero, V.K. Prasanna, S. Vajapeyam (Eds.), High Performance Computing—HiPC 2000. HiPC 2000, Lecture Notes in Computer Science, vol. 1970, Springer, Berlin, Heidelberg, 2000.

[10] W. Liao, J.M. Basile, L. He, in: Leakage power modeling and reduction with data retention, IEEE/ACM International Conference on Computer Aided Design, 2002. ICCAD 2002, 2002, pp. 714–719.

[11] H. Esmaeilzadeh, E. Blem, R.S. Amant, K. Sankaralingam, D. Burger, in: Dark silicon and the end of multicore scaling, Proceedings of the 38th Annual International Symposium on Computer Architecture (ISCA), San Jose, CA, 2011, pp. 365–376.

[12] N. Hardavellas, M. Ferdman, B. Falsafi, A. Ailamaki, Toward dark silicon in servers, IEEE Micro. 31 (4) (2011) 6–15.

[13] G.M. Amdahl, AFIPS Conference Proceedings, Validity of the single-processor approach to achieving large scale computing capabilities, vol. 30, AFIPS Press, Reston, Va., 1967, pp. 483–485. Atlantic City, N.J., Apr. 18-20.

[14] B.M. Rogers, A. Krishna, G.B. Bell, K. Vu, X. Jiang, Y. Solihin, in: Scaling the bandwidth wall: challenges in and avenues for CMP scaling, Proceedings of the 36th Annual International Symposium on Computer Architecture (ISCA '09). ACM, New York, NY, USA, 2009, pp. 371–382. https://doi.org/10.1145/1555754.1555801.

[15] D. Liu, C. Svensson, Trading speed for low-power by choice of supply and threshold voltages, IEEE J. Solid-State Circuits 28 (1993) 10–17.

[16] C. Kim, D. Burger, S.W. Keckler, in: An adaptive, non-uniform cache structure for wire-delay dominated on-chip cach- es, Proceedings of the 10th International Conference on Architectural Support for Programming Languages and Operating Systems, 2002.

[17] J. Huh, C. Kim, H. Shafi, L. Zhang, D. Burger, S.W. Keckler, in: A NUCA substrate for flexible CMP cache sharing, Proceedings of the 19th Annual International Conference on Supercomputing, 2005.

[18] B.M. Beckmann, M.R. Marty, D.A. Wood, in: ASR: adaptive selective replication for CMP caches, Proceedings of the 39th Annual IEEE/ACM International Symposium on Microarchitecture, 2006.

[19] J. Chang, G.S. Sohi, in: Cooperative caching for chip multiprocessors, Proceedings of the 33rd Annual International Symposium on Computer Architecture, 2006.

[20] Z. Chishti, M.D. Powell, T.N. Vijaykumar, in: Optimizing replication, communication, and capacity allocation in CMPs, Proceedings of the 32nd Annual International Symposium on Computer Architecture, 2005.

[21] M. Zhang, K. Asanovic, in: Victim replication: maximizing capacity while hiding wire delay in tiled chip multiprocessors, In Proceedings of the 32nd Annual International Symposium on Computer Architecture, 2005.

[22] Z. Guz, I. Keidar, A. Kolodny, U.C. Weiser, in: Utilizing shared data in chip multiprocessors with the Nahalal architec- ture, Proceedings of the 20th Annual ACM Symposium on Parallelism in Algorithms and Architectures, 2008.

[23] S. Cho, L. Jin, in: Managing distributed, shared L2 caches through OS-level page allocation, Proceedings of the 39th Annual IEEE/ACM International Symposium on Microarchitecture, 2006.

[24] N. Hardavellas, M. Ferdman, B. Falsafi, A. Ailamaki, in: Reactive NUCA: near-optimal block placement and replication in distributed caches, Proceedings of the 36th Annual International Symposium on Computer Architecture, 2009.

[25] B.M. Beckmann, D.A. Wood, in: Managing wire delay in large chip-multiprocessor caches, Proceedings of the 37th Annual IEEE/ACM International Symposium on Microarchitecture, 2004.

[26] M. Kandemir, F. Li, M.J. Irwin, S.W. Son, in: A novel migration-based NUCA design for chip multiprocessors, Proceedings of the 2008 ACM/IEEE Conference on Supercomputing, 2008.

[27] P. Ricci, S. Darius, D. Gebhardt, R. Balasubramonian, in: Leveraging bloom filters for smart search within NUCA caches, Proceedings of the 2006 Workshop on Complexity-Effective Design, 2006.

[28] M. Ferdman, A. Adileh, O. Kocberber, S. Volos, M. Alisafaee, D. Jevdjic, C. Kaynak, A.D. Popescu, A. Ailamaki, B. Falsafi, in: Clearing the clouds: a study of emerging scale-out workloads on modern hardware, Proceedings of the International Conference on Architectural Support for Programming Languages and Operating Systems, 2012.

[29] N. Hardavellas, I. Pandis, R. Johnson, N. Mancheril, A. Ailamaki, B. Falsafi, in: Database servers on chip multiprocessors: limitations and opportunities, The Conference on Innovative Data Systems Research, 2007.

[30] T. Kgil, S. D'Souza, A. Saidi, N. Binkert, R. Dreslinski, T. Mudge, S. Reinhardt, K. Flautner, in: PicoServer: using 3D stacking technology to enable a compact energy efficient chip multiprocessor, Proceedings of the International Conference on Architectural Support for Programming Languages and Operating Systems, 2006.

[31] P. Lotfi-Kamran, B. Grot, M. Ferdman, S. Volos, O. Kocberber, J. Picorel, A. Adileh, D. Jevdjic, S. Idgunji, E. Ozer, B. Falsafi, in: Scale-out processors, International Symposium on Computer Architecture, 2012, pp. 500–511.

[32] M. Ferdman, P. Lotfi-Kamran, K. Balet, B. Falsafi, in: Cuckoo directory: a scalable directory for many-core systems, Proceedings of the 17th IEEE International Symposium on High Performance Computer Architecture, 2011.

[33] L.A. Barroso, K. Gharachorloo, R. McNamara, A. Nowatzyk, S. Qadeer, B. Sano, S. Smith, R. Stets, B. Verghese, in: Piranha: a scalable architecture based on single-chip multiprocessing, ISCA '00: 27th Annual International Symposium on Computer Architecture, New York, NY, 2000.

[34] R. Golla, in: Niagara2: a highly threaded server-on-a-chip, Fall Microprocessor Forum 2006, San Jose, CA, 2006.

[35] A. Gupta, W. Weber, T. Mowry, in: Reducing memory and traffic requirements for scalable directory-based cache coherence schemes, ICPP '90: 1990 International Conference on Parallel Processing, Urbana-Champaign, IL, 1990.

[36] D. Sanchez, C. Kozyrakis, in: SCD: a scalable coherence directory with flexible sharer set encoding, Proceedings of the 18th International Symposium on High Performance Computer Architecture (HPCA), 2012.

[37] M.E. Acacio, J. González, J.M. García, J. Duato, in: A new scalable directory architecture for large-scale multiprocessors, HPCA '01: 7th International Symposium on High-Performance Computer Architecture, Washington, DC, 2001.

[38] A. Agarwal, R. Simoni, J. Hennessy, M. Horowitz, in: An evaluation of directory schemes for cache coherence, ISCA '88: 15th Annual International Symposium on Computer Architecture, Los Alamitos, CA, 1988.

[39] L.M. Censier, P. Feautrier, A new solution to coherence problems in Mulicache systems, IEEE Trans. Comput. 27 (12) (1978) 1112–1118.

[40] D. Chaiken, J. Kubiatowicz, A. Agarwal, in: Limit- LESS directories: a scalable cache coherence scheme, ASPLOS-IV: 4th International Conference on Architectural Support for Programming Languages and OS, New York, NY, 1991.

[41] G. Chen, in: SLiD—a cost-effective and scalable limited-directory scheme for cache coherence, PARLE '93: Parallel Architectures and Languages Europe, Heidelberg, Germany, 1993.

[42] J.H. Choi, K.H. Park, in: Segment directory enhancing the limited directory cache coherence schemes, IPPS '99/SPDP '99: 13th International Symposium on Parallel Processing and the 10th Symposium on Parallel and Distributed Processing, Washington, DC, 1999.

[43] C. Fensch, M. Cintra, in: An OS-based alternative to full hardware coherence on tiled CMPs, HPCA '08: 14th International Symposium on High Performance Computer Architecture, Salt Lake City, UT, 2008.

[44] M.M. Martin, M.D. Hill, in: Token coherence: decoupling performance and correctness, ISCA '03: 30th Annual International Symposium on Computer Architecture, New York, NY, 2003.

[45] A. Ros, M.E. Acacio, J.M. Garcia, in: DiCo-CMP: efficient cache coherency in tiled CMP architectures, IPDPS '08: 22nd International Parallel & Distributed Processing Symposium, Miami, FL, 2008.

[46] J. Zebchuk, V. Srinivasan, M.K. Qureshi, A. Moshovos, in: A tagless coherence directory, MICRO '09: 2009 42st International Symposium on Microarchitecture, New York, NY, 2009.

[47] S.L. Guo, H.X. Wang, Y.B. Xue, C.M. Li, D.S. Wang, Hierarchical cache directory for CMP, J. Comput. Sci. Technol. 25 (2) (2010) 246–256.

[48] A. Moshovos, G. Memik, B. Falsafi, A. Choudhary, in: JETTY: filtering snoops for reduced energy consumption in SMP servers, HPCA '01: Proceedings of the 7th International Symposium on High-Performance Computer Architecture, Washington, DC, USA, 2001.

[49] P. Lotfi-Kamran, M. Ferdman, D. Crisan, B. Falsafi, in: TurboTag: lookup filtering to reduce coherence directory power, International Symposium on Low Power Electronics and Design, 2010, pp. 377–382.

[50] J. Kim, J. Balfour, W. Dally, Flattened butterfly topology for on-chip networks, in: Proceedings of the 40th Annual IEEE/ACM International Symposium on Microarchitecture (MICRO), 2007, pp. 172–182.

[51] A. Jain, R. Parikh, V. Bertacco, High-radix on-chip networks with low-radix routers, in: Proceedings of the IEEE/ACM International Conference on Computer-Aided Design (ICCAD), 2014, pp. 289–294.

[52] B. Grot, J. Hestness, S.W. Keckler, O. Mutlu, in: Express cube topologies for on-chip interconnects, Proceedings of the 15th IEEE International Symposium on High-Performance Computer Architecture (HPCA), 2009, pp. 163–174.

[53] M.M. Kim, J.D. Davis, M. Oskin, T. Austin, in: Polymorphic on-chip networks, Proceedings of the 35th Annual International Symposium on Computer Architecture (ISCA), 2008, pp. 101–112.

[54] H. Yang, J. Tripathi, N.E. Jerger, D. Gibson, in: Dodec: random-link, low-radix on-chip networks, Proceedings of the 47th Annual IEEE/ACM International Symposium on Microarchitecture (MICRO), 2014, pp. 496–508.

[55] U.Y. Ogras, R. Marculescu, It's a small world after all: Noc performance optimization via long-range link insertion, IEEE Trans. Very Large Scale Integr. VLSI Syst. 14 (7) (2006) 693–706.

[56] J.C. Hu, R. Marculescu, in: DyAD—smart routing for networks-on-chip, Proceedings of the Design Automation Conference, 2004, pp. 260–263.

[57] M. Li, Q.-A. Zeng, W.-B. Jone, in: DyXY—a proximity congestion-aware deadlock-free dynamic routing method for network on chip, Proceedings of the Design Automation Conference, 2006, pp. 849–852.

[58] P. Gratz, B. Grot, S.W. Keckler, in: Regional congestion awareness for load balance in networks-on-Chip, Proceedings of the 14th IEEE International Symposium on High-Performance Computer Architecture (HPCA), 2008, pp. 203–214.

[59] P. Lotfi-Kamran, A.-M. Rahmani, M. Daneshtalab, A. Afzali- Kusha, Z. Navabi, EDXY—a low cost congestion- aware routing algorithm for network-on-chips, J. Syst. Archit. 56 (7) (2010) 256–264.

[60] S. Ma, N.E. Jerger, Z. Wang, in: DBAR: an efficient routing algorithm to support multiple concurrent applications in networks-on-chip, Proceedings of the 38th Annual International Symposium on Computer Architecture (ISCA), 2011, pp. 413–424.

[61] B. Fu, Y. Han, J. Ma, H. Li, X. Li, in: An Abacus turn model for time/space-efficient reconfigurable routing, Proceedings of the 38th Annual International Symposium on Computer Architecture (ISCA), 2011, pp. 259–270.

[62] P. Lotfi-Kamran, Per-packet global congestion estimation for fast packet delivery in networks-on-Chip, J. Supercomput. 71 (9) (2015) 3419–3439.

[63] P. Lotfi-Kamran, M. Modarressi, H. Sarbazi-Azad, in: Near-ideal networks-on-chip for servers, Proceedings of the 23rd IEEE International Symposium on High-Performance Computer Architecture (HPCA), 2017.

[64] R. Das, O. Mutlu, T. Moscibroda, C.R. Das, in: Argia: exploiting packet latency slack in on-chip networks, Proceedings of the 37th Annual International Symposium on Computer Architecture (ISCA), 2010, pp. 106–116.

[65] J.W. Lee, M.C. Ng, K. Asanovic, in: Globally synchronized frames for guaranteed quality-of-service in on-chip Networks, Proceedings of the 35th Annual International Symposium on Computer Architecture (ISCA), 2008, pp. 89–100.

[66] R. Das, O. Mutlu, T. Moscibroda, C.R. Das, in: Application-aware prioritization mechanisms for on-chip networks, Proceedings of the 42nd Annual IEEE/ACM International Symposium on Microarchitecture (MICRO), 2009, pp. 280–291.

[67] E. Bolotin, Z. Guz, I. Cidon, R. Ginosar, A. Kolodny, in: The power of priority: NoC based distributed cache coherency, Proceedings of the 1st IEEE/ACM International Symposium on Networks-on-Chip (NOCS), 2007, pp. 117–126.

[68] H. Matsutani, M. Koibuchi, H. Amano, T. Yoshinaga, in: Prediction router: yet another low latency on-chip router architecture, Proceedings of the 15th IEEE International Symposium on High-Performance Computer Architecture (HPCA), 2009, pp. 367–378.

[69] G. Michelogiannakis, N. Jiang, D. Becker, W.J. Dally, in: Packet chaining: efficient single-cycle allocation for on-chip networks, Proceedings of the 44th Annual IEEE/ACM International Symposium on Microarchitecture (MICRO), 2011, pp. 83–94.

[70] C.A. Nicopoulos, D. Park, J. Kim, N. Vijaykrishnan, M.S. Yousif, C.R. Das, in: ViChaR: a dynamic virtual channel regulator for network-on-chip routers, Proceedings of the 39th Annual IEEE/ACM International Symposium on Microarchitecture (MICRO), 2006, pp. 333–346.

[71] Y. Xu, B. Zhao, Y. Zhang, J. Yang, in: Simple virtual channel allocation for high throughput and high frequency on-chip routers, Proceeding of the 16th IEEE International Symposium on High-Performance Computer Architecture (HPCA), 2010, pp. 1–11.

[72] R. Mullins, A. West, S. Moore, in: Low-latency virtual-channel routers for on-chip networks, Proceedings of the 31st Annual International Symposium on Computer Architecture (ISCA), 2004, pp. 188–197.

[73] A. Kumar, L.-S. Peh, P. Kundu, N.K. Jha, in: Express virtual channels: towards the ideal interconnection fabric, Proceedings of the 34th Annual International Symposium on Computer Architecture (ISCA), 2007, pp. 150–161.

[74] A. Kumar, L.-S. Peh, N.K. Jha, in: Token flow control, Proceedings of the 41st Annual IEEE/ACM International Symposium on Microarchitecture (MICRO), 2008, pp. 342–353.

[75] M. Ahn, E.J. Kim, in: Pseudo-circuit: accelerating communication for on-chip interconnection networks, Proceedings of the 43rd Annual IEEE/ACM International Symposium on Microarchitecture (MICRO), 2010, pp. 399–408.

[76] L.-S. Peh, W.J. Dally, in: Flit-reservation flow control, Proceedings of the 6th IEEE International Symposium on High-Performance Computer Architecture (HPCA), 2000, pp. 73–84.

[77] T. Moscibroda, O. Mutlu, in: A case for bufferless routing in on-chip networks, Proceedings of the 36th Annual International Symposium on Computer Architecture (ISCA), 2009, pp. 196–207.

[78] M. Hayenga, N.E. Jerger, M. Lipasti, in: SCARAB: a single cycle adaptive routing and bufferless network, Proceedings of the 42nd Annual IEEE/ACM International Symposium on Microarchitecture (MICRO), 2009, pp. 244–254.

[79] A. Abousamra, A.K. Jones, R. Melhem, in: Proactive circuit allocation in multiplane NoCs, Proceedings of the 50th Annual Design Automation Conference (DAC), 2013, pp. 35:1–35:10.

[80] R.A. Stefan, A. Molnos, K. Goossens, dAElite: a TDM NoC supporting QoS, multicast, and fast connection set-up, IEEE Trans. Comput. 63 (3) (2014) 583–594.

[81] P. Lotfi-Kamran, M. Modarressi, H. Sarbazi-Azad, An efficient hybrid-switched network-on-chip for chip multi-processors, IEEE Trans. Comput. 65 (5) (2016) 1656–1662.

[82] P. Lotfi-Kamran, B. Grot, B. Falsafi, in: NOC-Out: microarchitecting a scale-out processor, Proceedings of the 45th Annual IEEE/ACM International Symposium on Microarchitecture (MICRO), 2012, pp. 177–187.

[83] D. Jevdjic, S. Volos, B. Falsafi, in: Die-stacked dram caches for servers: hit ratio, latency, or bandwidth? Have it all with footprint cache, Proceedings of the 40th Annual International Symposium on Computer Architecture, 2013.

[84] M.K. Qureshi, G.H. Loh, in: Fundamental latency trade-off in architecting DRAM caches: outperforming impractical SRAM-tags with a simple and practical design, Proceedings of the 45nd Annual IEEE/ACM International Symposium on Microarchitecture (MICRO), 2012, pp. 235–246.

[85] D. Jevdjic, G.H. Loh, C. Kaynak, B. Falsafi, in: Unison cache: a scalable and effective die-stacked DRAM cache, Proceedings of the 47nd Annual IEEE/ACM International Symposium on Microarchitecture (MICRO), 2014, pp. 25–37.

[86] X. Jiang, N. Madan, L. Zhao, M. Upton, R. Iyer, S. Makineni, D. Newell, Y. Solihin, R. Balasubramonian, CHOP: integrating DRAM caches for CMP server platforms, IEEE Micro. 31 (1) (2011) 99–108.

[87] X. Jiang, N. Madan, L. Zhao, M. Upton, R. Iyer, S. Makineni, D. Newell, Y. Solihin, R. Balasubramonian, in: CHOP: adaptive filter-based DRAM caching for CMP server platforms, HPCA—16 2010 The Sixteenth International Symposium on High-Performance Computer Architecture, 2010, pp. 1–12.

[88] G. Lohand, M.D. Hill, Supporting very large DRAM caches with compound-access scheduling and MissMap, IEEE Micro. 32 (3) (2012) 70–78.

[89] G.H. Loh, M.D. Hill, in: Efficiently enabling conventional block sizes for very large die-stacked DRAM caches, Proceedings of the 44th Annual IEEE/ACM International Symposium on Microarchitecture (MICRO), 2011, pp. 454–464.

[90] G.H. Loh, in: Extending the effectiveness of 3D-stacked DRAM caches with an adaptive multi-queue policy, Proceedings of the 42nd Annual IEEE/ACM International Symposium on Microarchitecture, 2009, pp. 201–212.

[91] G. Loh, M. Hill, in: Supporting very large caches with conventional block sizes, Proceedings of the 44th International Symposium on Microarchitecture, Porto Alegre, Brazil, 2011.

[92] L. Zhao, R. Iyer, R. Illikkal, D. Newell, in: Exploring DRAM cache architectures for AMP server platforms, 25th IEEE International Conference on Computer Design, 2007, pp. 55–62.

[93] Micron's Hybrid Memory Cube Earns High Praise in Next-Generation Supercomputer, Micron Technology, Inc. Available http://investors.micron.com/releasedetail.cfm?ReleaseID=805283.

[94] Y. Deng, W.P. Maly, 3-Dimensional VLSI: A 2.5-Dimensional Integration Scheme, first ed., Springer, Berlin Heidelberg, 2010.

[95] T. Johnson, U. Nawathe, in: An 8-core, 64-thread, 64-bit power efficient Sparc SoC (Niagara2), Proceedings of the 2007 International Symposium on Physical Design, 2007.

[96] Marvell, Marvell Unveils 1.6GHz Quad-Core ARMADA XP Platform for Enterprise Class Cloud Computing Applications (ONLINE) http://www.marvell.com/company/news/pressDetail.do?releaseID=1489, 2010.

[97] Cavium, ThunderX ARM Processors (ONLINE) http://www.cavium.com/ThunderX_ARM_Processors.html.

[98] K. Lim, P. Ranganathan, J. Chang, C. Patel, T. Mudge, S. Reinhardt, in: Understanding and designing new server architectures for emerging warehouse-computing environments, Proceedings of 35th Annual Internationa Symposium on Computer Architecture (ISCA 08), IEEE CS, 2008, pp. 315–326.

[99] S. Li, K. Lim, P. Faraboschi, J. Chang, P. Ranganathan, N. Jouppi, in: System-level integrated server architectures for scale-out datacenters, Proceedings of the 44th Annual International Symposium on Microarchitecture, 2011.

[100] D.G. Andersen, et al., in: FAWN: a fast array of wimpy nodes, Proceedings of the 22nd ACM SIGOPS Symposium on Operating Systems Principles, 2009, pp. 1–14.

[101] A.M. Caulfield, L.M. Grupp, S. Swanson, in: Gordon: using flash memory to build fast, power-efficient clusters for data-intensive applications, 14th International Conference on Architectural Support for Programming Languages and Operating Systems, 2009.

[102] E. Ozer, K. Flautner, S. Idgunji, A. Saidi, Y. Sazeides, B. Ahsan, N. Ladas, C. Nicopoulos, I. Sideris, B. Falsafi, A. Adileh, M. Ferdman, P. Lotfi-Kamran, M. Kuulusa, P. Marchal, N. Minas, in: EuroCloud: energy-conscious 3D server-on-chip for green cloud services, Workshop on Architectural Concerns in Large Datacenters in Conjunction with International Symposium on Computer Architecture (ISCA), 2010.

[103] B. Grot, D. Hardy, P. Lotfi-Kamran, B. Falsafi, C. Nicopoulos, Y. Sazeides, Optimizing data-center TCO with scale-out processors, IEEE Micro. 32 (5) (2012) 52–63. Special Issue on Energy-Aware Computing.

[104] R. Hameed, W. Qadeer, M. Wachs, O. Azizi, A. Solomatnikov, B.C. Lee, S. Richardson, C. Kozyrakis, M. Horowitz, in: Understanding sources of inefficiency in general-purpose chips, Proceedings of International Symposium on Computer Architecture, 2010.

[105] V. Govindaraju, C.-H. Ho, K. Sankaralingam, in: Dynamically specialized datapaths for energy efficient computing, Proceedings of the 17th Annual International Symposium on High Performance Computer Architecture, 2011.

[106] S. Gupta, S. Feng, A. Ansari, S. Mahlke, D. August, in: Bundled execution of recurring traces for energy-efficient general purpose processing, Proceedings of the 44th Annual IEEE/ACM International Symposium on Microarchitecture, 2011.

[107] J. Sampson, G. Venkatesh, N. Goulding-Hotta, S. Garcia, S. Swanson, M. Taylor, in: Efficient complex operators for irregular codes, Proceedings of the 17th Annual International Symposium on High Performance Computer Architecture, 2011.

[108] G. Venkatesh, J. Sampson, N. Goulding-Hotta, S.K. Venkata, M.B. Taylor, S.S. QsCores, in: Trading dark silicon for scalable energy efficiency with quasi-specific cores, Proceedings of the 44th Annual IEEE/ACM International Symposium on Microarchitecture, 2011.

[109] O. Kocberber, B. Grot, J. Picorel, B. Falsafi, K. Lim, P. Ranganathan, in: Meet the walkers: accelerating index traversals for in-memory databases, Proceedings of the

46th Annual IEEE/ACM International Symposium on Microarchitecture, MICRO-46, 2013, pp. 468–479.

[110] T.J. Ham, L. Wu, N. Sundaram, N. Satish, M. Martonosi, in: Graphicionado: a high-performance and energy-efficient accelerator for graph analytics, 2016 49th Annual IEEE/ACM International Symposium on Microarchitecture (MICRO), Taipei, 2016, pp. 1–13.

[111] J. Ahn, S. Hong, S. Yoo, O. Mutlu, K. Choi, in: A scalable processing-in-memory accelerator for parallel graph processing, 2015 ACM/IEEE 42nd Annual International Symposium on Computer Architecture (ISCA), Portland, OR, 2015, pp. 105–117.

[112] I. Magaki, M. Khazraee, L.V. Gutierrez, M.B. Taylor, in: ASIC clouds: specializing the datacenter, 2016 ACM/IEEE 43rd Annual International Symposium on Computer Architecture (ISCA), Seoul, 2016, pp. 178–190.

[113] D. Koeplinger, R. Prabhakar, Y. Zhang, C. Delimitrou, C. Kozyrakis, K. Olukotun, in: Automatic generation of efficient accelerators for reconfigurable hardware, 2016 ACM/IEEE 43rd Annual International Symposium on Computer Architecture (ISCA), Seoul, 2016, pp. 115–127.

[114] V.W. Lee, C. Kim, J. Chhugani, M. Deisher, D. Kim, A.D. Nguyen, N. Satish, M. Smelyanskiy, S. Chennupaty, P. Hammarlund, R. Singhal, P. Dubey, in: Debunking the 100X GPU vs. CPU myth: an evaluation of throughput computing on CPU and GPU, Proceedings of the 37th annual international symposium on Computer architecture (ISCA '10), ACM, New York, NY, USA, 2010, pp. 451–460.

[115] C. Gregg, K. Hazelwood, in: Where is the data? Why you cannot debate CPU vs. GPU performance without the answer, (IEEE ISPASS) IEEE International Symposium on Performance Analysis of Systems and Software, Austin, TX, 2011, pp. 134–144.

[116] S. Mittal, J.S. Vetter, A Survey of CPU-GPU heterogeneous computing techniques, ACM Comput. Surv. 47 (4) (2015). 69 (35 pages).

[117] E.S. Chung, P.A. Milder, J.C. Hoe, K. Mai, in: Single-chip heterogeneous computing: does the future include custom logic, FPGAs, and GPGPUs?, 2010 43rd Annual IEEE/ACM International Symposium on Microarchitecture, Atlanta, GA, 2010, pp. 225–236.

[118] A. Putnam, A.M. Caulfield, E.S. Chung, D. Chiou, K. Constantinides, J. Demme, H. Esmaeilzadeh, J. Fowers, G.P. Gopal, J. Gray, M. Haselman, S. Hauck, S. Heil, A. Hormati, J.-Y. Kim, S. Lanka, J. Larus, E. Peterson, S. Pope, A. Smith, J. Thong, P.Y. Xiao, D. Burger, in: A reconfigurable fabric for accelerating large-scale datacenter services, Proceeding of the 41st Annual International Symposium on Computer Architecuture (ISCA '14). IEEE Press, Piscataway, NJ, USA, 2014, pp. 13–24.

[119] M.F. Deering, S.A. Schlapp, M.G. Lavelle, in: FBRAM:A new form of memory optimized for 3D graphics, Proceedings of the 21st annual conference on Computer graphics and interactive techniques SIGGRAPH '94, 1994.

[120] J. Draper, J. Chame, M. Hall, C. Steele, T. Barrett, J. LaCoss, J. Granacki, J. Shin, C. Chen, C.W. Kang, I. Kim, G. Daglikoca, in: The architecture of the DIVA processing-in-memory chip, Proceedings of the 16th International Conference on Supercomputing, ICS '02, 2002.

[121] D. Patterson, T. Anderson, N. Cardwell, R. Fromm, K. Keeton, C. Kozyrakis, R. Thomas, K. Yelick, A Case for Intelligent RAM, IEEE Micro. 17 (2) (1997).

[122] D.G. Elliott, W.M. Snelgrove, M. Stumm, Computational RAM: a memory-SIMD hybrid and its application to DSP, in: Custom Integrated Circuits Conferencevol. 30, 1992.

[123] Y. Kang, W. Huang, S.-M. Yoo, D. Keen, Z. Ge, V. Lam, P. Pattnaik, J. Torrellas, in: FlexRAM: toward an advanced intelligent memory system, Proceedings of the 1999 IEEE International Conference on Computer Design, ICCD '99, 2012.

[124] K. Mai, T. Paaske, N. Jayasena, R. Ho, W.J. Dally, M. Horowitz, in: Smart memories: a modular reconfigurable architecture, International Symposium on Computer Architecture, 2000.

[125] M. Oskin, F. Chong, T. Sherwood, Active pages: a computation model for intelligent memory, in: International Symposium on Computer Architecture, 1998.

[126] A. Farmahini-Farahani, J.H. Ahn, K. Morrow, N.S. Kim, in: NDA: near-DRAM acceleration architecture leveraging commodity DRAM devices and standard memory modules, High Performance Computer Architecture (HPCA), 2015 IEEE 21st International Symposium, 2015.

[127] H. Asghari-Moghaddam, Y. HoonSon, J. HoAhn, N. SungKim, in: Chameleon: versatile and practical near-DRAM acceleration architecture for large memory systems, 2016 49th Annual IEEE/ACM International Symposium on Microarchitecture (MICRO), 2016.

[128] K. Hsieh, E. Ebrahimi, G. Kim, N. Chatterjee, M. O'Connor, N. Vijaykumar, O. Mutlu, S.W. Keckler, in: Transparent offloading and mapping (TOM): enabling programmer-transparent near-data processing in GPU systems, Proceedings of the 43rd International Symposium on Computer Architecture (ISCA 2016), 2016.

[129] M. Gao, J. Pu, X. Yang, M. Horowitz, C. Kozyrakis, TETRIS: scalable and efficient neural network acceleration with 3D memory, Proceedings of the 22nd International Conference on Architectural Support for Programming Languages and Operating Systems (ASPLOS), 2017.

[130] L. Nai, R. Hadidi, J. Sim, H. Kim, P. Kumar, H. Kim, in: GraphPIM: enabling instruction-level PIM offloading in graph computing frameworks, 2017 IEEE 23rd International Symposium on High. Performance Computer Architecture (HPCA), 2017.

[131] D. Zhang, N. Jayasena, A. Lyashevsky, J. Greathouse, L. Xu, M. Ignatowski, in: TOP-PIM: throughput-oriented programmable processing in memory, Proceedings of International Symposium on High-performance Parallel and Distributed Computing (HPDC), 2014.

[132] S. Pugsley, J. Jestes, R. Balasubramonian, V. Srinivasan, A. Buyuktosunoglu, A. Davis, F. Li, Comparing implementations of near-data computing with in-memory MapReduce workloads, IEEE Micro. 34 (4) (2014).

[133] R. Nair, S. Antao, C. Bertolli, P. Bose, J. Brunheroto, T. Chen, C. Cher, C. Costa, J. Doi, C. Evangelinos, B. Fleischer, T. Fox, D. Gallo, L. Grinberg, J. Gunnels, A. Jacob, P. Jacob, H. Jacobson, T. Karkhanis, C. Kim, J. Moreno, J. O'Brien, M. Ohmacht, Y. Park, D. Prener, B. Rosenburg, K. Ryu, O. Sallenave, M. Serrano, P. Siegl, K. Sugavanam, Z. Sura, Active memory cube: a processing-in-memory architecture for exascale systems, IBM J. Res. Dev. 59 (2/3) (2015) 17:1–17:14.

[134] Q. Guo, N. Alachiotis, B. Akin, F. Sadi, G. Xu, T.-M. Low, L. Pileggi, J. Hoe, F. Franchetti, in: 3D-stacked memory-side acceleration: accelerator and system design, Workshop on Near-Data Processing (WoNDP), 2014.

[135] C. Shelor, K. Kavi, S. Adavally, in: Dataflow based near data processing using coarse grain reconfigurable logic, Workshop on Near-Data Processing (WoNDP), 2015.

[136] Q. Zhu, T. Graf, H. Sumbul, L. Pileggi, F. Franchetti, in: Accelerating sparse matrix-matrix multiplication with 3D-stacked logic-in-memory hardware, Proceedings of the IEEE High Performance Extreme Computing Conference (HPEC), 2013.

[137] M. Gao, C. Kozyrakis, in: HRL: efficient and flexible reconfigurable logic for near-data processing, 22nd IEEE Symposium on High Performance Computer Architecture (HPCA), 2016.

[138] A. Farmahini-Farahani, J.H. Ahn, K. Morrow, N.S. Kim, DRAMA: an architecture for accelerated processing near memory, CAL 14 (1) (2015) 26–29.

[139] D. Kim, J. Kung, S. Chai, S. Yalamanchili, S. Mukhopadhyay, in: NeuroCube: a programmable digital neuromorphic architecture with high-density 3D memory, 2016

ACM/IEEE 43rd Annual International Symposium on Computer Architecture (ISCA), 2016.

[140] M. Samadi, J. Lee, D.A. Jamshidi, A. Hormati, S. Mahlke, in: SAGE: self-tuning approximation for graphics engines, Proceedings of the 46th Annual IEEE/ACM International Symposium on Microarchitecture, MICRO-46, Association for Computing Machinery, 2013.

[141] M. Samadi, D.A. Jamshidi, J. Lee, S. Mahlke, in: Paraprox: pattern-based approximation for data parallel applications, Proceedings ASPLOS, 2014.

[142] J.-M. Arnau, J.-M. Parcerisa, P. Xekalakis, in: Eliminating redundant fragment shader executions on a mobile gpu via hardware memoization, 2014 ACM/IEEE 41st International Symposium on Computer Architecture (ISCA), 2014.

[143] J. Sartori, R. Kumar, Branch and data herding: reducing control and memory divergence for error-tolerant GPU applications, IEEE Trans. Multimedia 15 (2) (2013) 279–290.

[144] S. Liu, K. Pattabiraman, T. Moscibroda, B.G. Zorn, in: Flikker: saving refresh-power in mobile devices through critical data partitioning, Proceedings ASPLOS, 2011.

[145] A. Sampson, J. Nelson, K. Strauss, L. Ceze, in: Approximate storage in solid-state memories, Proceedings of the International Symposium on Microarchitecture (MICRO '46), 2013.

[146] H. Esmaeilzadeh, A. Sampson, L. Ceze, D. Burger, in: Architecture support for disciplined approximate programming, Proceedings ASPLOS, 2012.

[147] L.N. Chakrapani, B.E.S. Akgul, S. Cheemalavagu, P. Korkmaz, K.V. Palem, B. Seshasayee, in: Ultra-efficient (embedded) SOC architectures based on probabilistic CMOS (PCMOS) technology, DATE '06 Proceedings of the conference on Design, automation and test in Europe, 2006.

[148] L. Leem, H. Cho, J. Bau, Q.A. Jacobson, S. Mitra, in: ERSA: error resilient system architecture for probabilistic applications, Design, Automation & Test in Europe Conference & Exhibition (DATE), 2010.

[149] S. Sidiroglou-Douskos, S. Misailovic, H. Hoffmann, M. Rinard, in: Managing performance vs. accuracy trade-offs with loop perforation, Proceedings of 19th ACM SIGSOFT Symposium and the 13th European Conference on Foundations of Software Engineering, ESEC/FSE, 2011.

[150] S. Misailovic, S. Sidiroglou, H. Hoffman, M. Rinard, in: Quality of service profiling, Proceedings of ICSE, 2010.

[151] M. Rinard, H. Hoffmann, S. Misailovic, S. Sidiroglou, in: Patterns and statistical analysis for understanding reduced resource computing, OOPSLA'10 - Proceedings of the 2010 ACM SIGPLAN Conference on Object Oriented Programming, Systems, Languages, and Applications, 2010.

[152] W. Baek, T.M. Chilimbi, in: Green: a framework for supporting energy-conscious programming using controlled approximation, Proceedings of the 31st ACM SIGPLAN conference on programming language design and implementation (PLDI), 2010.

[153] J. Ansel, C. Chan, Y.L. Wong, M. Olszewski, Q. Zhao, A. Edel- man, S. Amarasinghe, in: Petabricks: a language and compiler for algorithmic choice, Proceedings of the 2009 ACM SIGPLAN Conference on Programming Language Design and Implementation (PLDI), 2009.

[154] C. Alvarez, J. Corbal, M. Valero, Fuzzy memoization for floating-point multimedia applications, IEEE Trans. Comput. 54 (7) (2005) 922–927.

[155] M. de Kruijf, S. Nomura, K. Sankaralingam, in: Relax: an architectural framework for software recovery of hardware faults, Proceedings of the 37th Annual International Symposium on Computer Architecture, ISCA 2010, 2010.

[156] X. Li, D. Yeung, in: Application-level correctness and its impact on fault tolerance, IEEE 13th International Symposium on High Performance Computer Architecture (HPCA), 2007.

[157] X. Li, D. Yeung, Exploiting Application-Level Correctness for Low-Cost Fault Tolerance, Journal Instruction-Level Parallelism, 2008.

[158] M. de Kruijf, K. Sankaralingam, Exploring the synergy of emerging workloads and silicon reliability trends, Silicon Errors in Logic System Effects Workshop, 2009.

[159] Y. Fang, H. Li, X. Li, in: A fault criticality evaluation framework of digital systems for error tolerant video applications, Proceedings of the 2011 Asian Test Symposium (ATS), 2011.

[160] A. Sampson, W. Dietl, E. Fortuna, D. Gnanapragasam, L. Ceze, D. Grossman, in: EnerJ: approximate data types for safe and general low-power computation, PLDI '11 Proceedings of the 32nd ACM SIGPLAN Conference on Programming Language Design and Implementation, 2011.

[161] S. Venkataramani, V.K. Chippa, S.T. Chakradhar, K. Roy, A. Raghunathan, in: Quality programmable vector processors for approximate computing, Proceedings of the 46th Annual. IEEE/ACM International Symposium on Microarchitecture (MICRO-46), 2013.

[162] A. Yazdanbakhsh, D. Mahajan, B. Thwaites, J. Park, A. Nagendrakumar, S. Sethuraman, K. Ramkrishnan, N. Ravindran, R. Jariwala, A. Rahimi, H. Esmaeilzadeh, K. Bazargan, in: Axilog: language support for approximate hardware design, Proceedings of the 2015 Design, Automation and Test in Europe Conference and Exhibition, DATE, 2015.

[163] A. Ranjan, A. Raha, S. Venkataramani, K. Roy, A. Raghunathan, ASLAN: synthesis of approximate sequential circuits, in: Design, Automation and Test in Europe Conference and Exhibition(DATE), Institute of Electrical and Electronics Engineers (IEEE), 2014.

[164] S. Venkataramani, A. Sabne, V. Kozhikkottu, K. Roy, A. Raghunathan, in: SALSA: systematic logic synthesis of approximate circuits, 49th Annual Design Automation Conference 2012, DAC'12, 2012.

[165] J. Miao, A. Gerstlauer, M. Orshansky, in: Approximate logic synthesis under general error magnitude and frequency constraints, ICCAD'13: IEEE/ACM International Conference on ComputerAided Design, 2013.

[166] K. Nepal, Y. Li, R.I. Bahar, S. Reda, in: ABACUS: a technique for automated behavioral synthesis of approximate computing circuits, Proceedings of the Conference on Design, Automation and Test in Europe, EDA Consortium, DATE'14, 2014.

[167] A. Lingamneni, C. Enz, K. Palem, C. Piguet, Synthesizing parsimonious inexact circuits through probabilistic design techniques, ACM Trans. Embed. Comput. Syst. 12 (2s) (2013) 93:1–93:26.

[168] A. Lingamneni, K.K. Muntimadugu, C. Enz, R.M. Karp, K.V. Palem, C. Piguet, in: Algorithmic methodologies for ultra-efficient inexact architectures for sustaining technology scaling, Proceedings of the 9th Conference on Computing Frontiers (CF'12), 2012.

[169] H. Esmaeilzadeh, A. Sampson, L. Ceze, D. Burger, in: Neural acceleration for general-purpose approximate programs, Proceedings of the 2012 45th Annual IEEE/ACM International Symposium on Microarchitecture, 2012.

[170] R.S. Amant, A. Yazdanbakhsh, J. Park, B. Thwaites, H. Esmaeilzadeh, A. Hassibi, L. Ceze, D. Burger, in: General-purpose code acceleration with limited-precision analog computation, Proceedings—International Symposium on Computer Architecture, 2014.

[171] B. Grigorian, N. Farahpour, G. Reinman, in: BRAINIAC: bringing reliable accuracy into neurally-implemented approximate computing, International Symposium on High-Performance Computer Architecture (HPCA), 2015.

[172] T. Moreau, M. Wyse, J. Nelson, A. Sampson, H. Esmaeilzadeh, L. Ceze, M. Oskin, in: SNNAP: approximate computing on programmable socs via neural acceleration, International Symposium on High-Performance Computer Architecture (HPCA), 2015.

[173] L. McAfee, K. Olukotun, in: EMEURO: a framework for generating multi-purpose accelerators via deep learning, Proceedings of the 13th Annual IEEE/ACM International Symposium on Code Generation and Optimization (CGO), 2015.

[174] B. Grigorian, G. Reinman, in: Accelerating divergent applications on SIMD architectures using neural networks, IEEE International Conference on Computer Design (ICCD), 2014.

[175] A. Yazdanbakhsh, J. Park, H. Sharma, P. Lotfi-Kamran, H. Esmaeilzadeh, in: Neural acceleration for GPU throughput processors, Proceedings of the 48th International Symposium on Microarchitecture, MICRO-48, 2015.

[176] A. Yazdanbakhsh, D. Mahajan, H. Esmaeilzadeh, P. Lotfi-Kamran, AxBench: a multiplatform benchmark suite for approximate computing, IEEE Des. Test 34 (2) (2017) 60–68.

ABOUT THE AUTHORS

Pejman Lotfi-Kamran is an associate professor of computer science and the manager of Iran's national compute-grid project at Institute for Research in Fundamental Sciences (IPM). His research interests include computer architecture, computer systems, approximate computing, and cloud computing. His recent work on scale-out server processor design lays the foundation for Cavium ThunderX. He has a PhD in computer science from the École Polytechnique Fédérale de Lausanne (EPFL). He received his MS and BS in computer engineering from the University of Tehran. He is a member of IEEE and the ACM.

 **Hamid Sarbazi-Azad** received the BSc degree in electrical and computer engineering from Shahid-Beheshti University, Tehran, Iran, in 1992, the MSc degree in computer engineering from the Sharif University of Technology, Tehran, Iran, in 1994, and the PhD degree in computing science from the University of Glasgow, Glasgow, United Kingdom, in 2002. He is currently professor of computer engineering at the Sharif University of Technology and heads the School of Computer Science, Institute for Research in Fundamental Sciences (IPM), Tehran, Iran. His research interests include high-performance computer/memory architectures, NoCs and SoCs, parallel and distributed systems, performance modeling/evaluation, and storage systems, on which he has published more than 300 refereed conference and journal papers. He received Khwarizmi International Award in 2006, TWAS Young Scientist Award in engineering sciences in 2007, and Sharif University Distinguished Researcher awards in years 2004, 2007, 2008, 2010, and 2013. He is now an associate editor of ACM Computing Surveys, Elsevier Computers and Electrical Engineering, and CSI Journal on Computer Science and Engineering.

CHAPTER TWO

Revisiting Processor Allocation and Application Mapping in Future CMPs in Dark Silicon Era

Mohaddeseh Hoveida*, Fatemeh Aghaaliakbari*, Majid Jalili*, Ramin Bashizade*, Mohammad Arjomand[†], Hamid Sarbazi-Azad*
*Department of Computer Engineering, Sharif University of Technology, Tehran, Iran
[†]School of Electrical and Computer Engineering, Georgia Institute of Technology, Atlanta, GA, United States

Contents

Abstract

With technology advances and the emergence of new fabrication and VLSI technologies, current and future chip multiprocessors (CMPs) are expected to have tens to hundreds of processing elements and Gigabytes of on-chip caches, which are connected by a high bandwidth network-on-chip (NoC). Unfortunately, due to limited power budget of a computing system, specially for its processing element(s), it is impossible to keep all cores, caches, and network elements working at highest voltage level—that would resulted in dark silicon computing era, where by employing system-level

Advances in Computers, Volume 110
ISSN 0065-2458
https://doi.org/10.1016/bs.adcom.2018.04.001

or architecture-level techniques, one can keep a great portion of a CMP elements OFF (or in dim mode) to meet the power budget of the system while the system still delivers a high-performance computation.

In this work, we first describe the importance of NoC design and management in delivering high-performance computation in a dark silicon-based CMP platform—we propose a novel highly scalable NoC architecture and its required management policies in order to support turning some routers/links/buffers OFF while guaranteeing that it delivers the bandwidth needs of the running application(s). Then, by employing the introduced NoC architecture, we propose to revisit the processor allocation strategy and application-to-core mapping algorithm in order to make maximum use of the provided NoC bandwidth and capability while meeting the power and performance goals of the hardware platform and application, respectively. Our extensive simulation results of a 64-core CMP model show that the proposed algorithms are able to improve the system performance by 10%–50% when running multithreaded applications.

1. INTRODUCTION

With the advances in CMOS fabrication technology and the emergence of new technologies (such as 3D chips and non-volatile memories), it is expected that future chip multiprocessors (CMPs) integrate hundreds of processing elements (or cores) and Gigabytes of on-chip cache memories. The large cache memory is largely banked in these systems and these is a high bandwidth and fast interconnection network (known as network-on-chip or NoC) between the cache banks and cores. Unfortunately, due to power budget of every computing system and especially its processor, one cannot make maximum use of this huge computation and memory capability at each time instant—Just a portion of all elements are allowed to work at high voltage level and the rest parts have to be either turned off or kept in dim mode (i.e., a low-voltage mode).

Compared to the conventional computing systems, such a new computing platform in *dark silicon era* encounters more serious power and temperature limitations. Indeed, resource allocation and management are very critical for such a system, because it determines (i) the system performance as well as the communication latency between cores and cache banks, and (ii) the thermal distribution of the chip. In other words, *a design goal can be how much one can accelerate applications' execution times on these systems such that overall power budget is met and the resulting temperature over all resources evenly distributes.*

While prior work have extensively study different optimization aspects of a CMP in dark silicon-based computing platforms (including optimizations at programming languages, compiler, architecture, and system-level), least attention is paid to the efficiency of underlying NoC as well as the processor allocation and application mapping in the tested CMP when considering the capability and limitations of the NoC. This is the focus of our research in current work.

In this chapter, we qualitatively describe the impact of the underlying NoC and its management policies on performance, power, and thermal efficiency of a CMP with tens of cores. As an NoC in such system has to deliver high bandwidth for the running application(s), one has to have a capability to turn off network resources on-demand such that overall power is reduced while still network is fully connected (or at least provides some connections between parts that are in ON mode). Having these goals in mind, we propose a novel NoC architecture, which is inspired by high-connectivity that C-Mesh topology gives. The proposed architecture has the capability of (i) sharing the bandwidth between multiple flows in the system and (ii) turning off the network resources at finest granularity (routers and buffers).

Having the performance and power management capabilities that the proposed NoC architecture gives, we revisit the "resource allocation" and "application-to-core mapping" algorithms such that limitations of system is met while the performance is still high. More accurately, this paper has the following two major contributions for resource management in NoCs:

1. We propose a processor allocation scheme which is aware of both performance and power, called Dark Silicon Aware Processor Allocation (DSAPA). Our evaluations show that the proposed solution improves performance by up to 38.7% (on average 15.5%) and reduce power consumption by up to 29.4% (on average 17.6%), respectively.

2. We also present two dark silicon aware mapping algorithms based on the proposed platform, one with High Capability of Power Saving (*HCPS*), and another with High Capability of Resource Sharing (*HCRS*). In both algorithms, we assume that the number of cores in the considered region for mapping is not necessarily equal to the number of tasks to be mapped. The main goal of these algorithms is to reduce the average execution time while meeting power constraints. More precisely, *HCPS* tries to best utilize all resources of the clusters hosting the mapped tasks. In this way, some unused clusters can be either powered off or assigned to other applications. So, this algorithm targets increasing the performance and power budget while reducing power consumption of NoC components. On the other hand, *HCRS* aims to improve the performance by

employing cores of different clusters and utilizing the resources of the unused cores in each cluster by active cores. Although this algorithm does not aim to have some clusters switched off, it satisfies power budget limits. Our evaluations show that *HCPS* and *HCRS* algorithms improve the average execution time by up to 28.6% and 39.2%, reduce NoC power consumption by up to 11.1% and 10%, total power consumption by up to 22.9% and 26.2%, and gain excess power budget up to 7.6% and 13.4%, respectively, with respect to the baseline architecture.

The rest of the chapter is organized as follows. Section 2 reviews main contributions of the related work. Section 3 introduces and evaluates the proposed NoC architecture (named SCMesh). The proposed processor allocation strategy and mapping algorithm are given in Sections 4 and 5, respectively. Section 6 describes the experimental setup and reports the results. Finally, Section 7 concludes the chapter.

2. RELATED WORK

With the rising trend of the number of available transistors on the chip and reducing parts of the chip that can be used simultaneously, previous works have largely been presented based on specialized processors in order to provide significant energy efficiency, compared to the general-purpose processors [1–3].

Unlike prior approaches that require many pipeline components to remain active, completely switching off the general-purpose processor pipeline in [2] provides higher energy savings. Goulding-Hotta et al. [1] offer an architecture for dark silicon age that uses conversation cores [3] that are specialized processors focusing on reducing energy and energy–delay product instead of increasing the performance. This architecture drastically reduces the energy of an Android-based mobile software stack. In [4], a framework for asymmetric multicores architectures is presented as a promising alternative in dark silicon era. In this framework, only a fraction of cores on the chip can be powered on to achieve optimal power-performance efficiency and keep the thermal design power budget intact.

Since IP mapping is an effective way to optimize the chip parameters by selecting a set of cores for running tasks, we studied the various heuristic algorithms proposed in recent years. The main aim of these algorithms is to improve performance, congestion, power consumption and throughput of the system. However, algorithms that target the performance improvement, mostly use contiguous mapping [5–7]. So they do not focus on power budget problem in dark silicon era. The mapping heuristics that are proposed in [8], map the communicating tasks of an application close to each other to minimize the communication overhead and improve performance. *NMAP* is an efficient algorithm [6] that maps tasks onto nodes with minimum path routing in the mesh topology which considers bandwidth and minimizes average communication delay, but do not concentrate on dark silicon issue. To address communication energy and performance issues, [9] presented an efficient technique for incremental runtime application mapping onto a homogeneous cluster-based NoC with multiple voltage levels. Besides, an energy-aware heuristic for dynamic task mapping with the aim of reducing energy consumption has been presented in [10]. In [11], an efficient power-aware template-based mapping algorithm for NoC is proposed in order to generate good mapping solutions with lower runtime under bandwidth and latency constraints. Aforementioned mapping strategies do not consider fixed upper bound on power consumption.

Authors in [12–14] presented methods for thermal management and hotspots reduction. However, they do not contrive to make better use of power budget for the sake of mitigating dark silicon problems. In [12], on-chip power consumption is distributed in order to reduce the maximum chip temperature. This method also corresponds to trade-off between network latency and temperature, but does not target performance aspect of dark silicon. A multiobjective ant colony algorithm is applied to map the IPs onto clusters in 2D mesh NoC with the aim of optimizing the total energy consumption and providing the thermal stability [13]. In [14], an application specific mapping algorithm is explored that maps IP cores onto the mesh topology with the purpose of decreasing peak temperature and communication cost by taking thermal correlation into account.

In NoC domain, Zhan et al. [15] mitigated dark silicon problems by presenting an efficient network power management method. It allows a chip to activate dark cores for throughput improvement and investigates routing support and thermal-aware floor planning to explore fine-grained sprinting process. In [16], a power management dark silicon aware platform for many-core systems is proposed to manage power by changing voltage and frequency of processing elements into a proper level. It employs a feedback controller to provide safe upper bound on power consumption of the whole system. A dark silicon aware runtime application mapping approach is proposed in [17]. The main idea is to gain higher power budget through a dark silicon patterning approach that locates active cores alongside the inactive ones. Thus, application tasks are distributed across the chip and balance heat distribution. As a result of the gained surplus budget, more cores at lower frequencies could be used and then make better resources utilization.

So far, on-chip interconnects in dark silicon era have received insufficient attention. In [18] a power-efficient NoC, called DimNoC, is proposed in which pure SRAM-based buffers in NoC routers are replaced with hybrid buffers through leveraging both drowsy SRAM and STT-RAM technologies. Therefore, it can use advantages of both memory technologies to achieve energy saving and area reduction. A novel fine-grained temperature-aware NoC-based MCSoC architecture named Shift Sprinting is proposed in [19] that handles high-performance QoS-aware mobile demands and gains high-reliability and performance by employing the concept of distributional sprinting. NoC not only directly influences the overall multicore systems performance, but also consumes a significant portion of the total chip power [12]. Therefore, we target the dark silicon problems in our research through a fine-grained power management architecture called *SCmesh*.

As previously stated, our study is limited to mapping the IP cores onto a regular cluster-based NoC architecture and mapping is subjected to two main challenges in dark silicon era, the performance and power constraints.

Future CMPs with tens to hundreds of processing cores need a scalable and high bandwidth intercore interconnection medium to guarantee high-performance and low-cost (low-power) design. Among different applicable communication topologies, such as network-on-chip (NoC) and bus and point-to-point, NoC is the only scalable option which has been achieved lots of researches during last decade [20–23]. When talking about CMPs and NoCs in dark silicon area, some conventional constraints on NoC design become very crucial. For instance, as Zhu et al. [12] have experimentally showed that NoC routers and cores have more power-to-area ratio compared to other on-chip components, which necessitates any NoC-based CMP design to be more low-power and highly scalable. Therefore, reducing the power of NoC components and turning off inactive cores can result in lower potential hotspots in each cluster. Thus, we focus on both router components and cores in our proposed platform. On the other hand, dark and active cores should be located in such a way that higher power budgets can be obtained. This is the main objective of our study.

Up to 75% impact of interconnection on overall performance introduces it as a key component in memory hierarchy [24]. In other words, studying large-scale CMPs without considering NoC role is inaccurate. Some prior researches including application specific approaches [25, 26] introduced special designs for power efficiency while maintaining performance; however, they do not consider NoC power. Some works, like [27] and [28], consider NoC as an important component in dark silicon design. Zhan et al. [27] presented an approach based on router power gating for short-burst computations using a specific routing. In [29] and [30], a voltage–frequency islands (VFIs) technique on CMPs was introduced which minimizes energy consumption under expected performance. Khdr et al. [31] select the active cores according to their voltage/frequency level with the goal of maximizing the performance. Obtaining this approach in large-scale CMPs means dim silicon (under clocked resources), while in our paper we focus on dark silicon context. A number of recent papers have focused on task migration over DVFS-based NoCs because of thermal constraints [32, 33]. In [34] and [35], task migration is employed for balancing temperature among the chip that generates less hotspots.

Due to the thermal hotspots, high temperature gradients and limited power budget in many-core systems, unused resources are going to be increased in the near future. Thus, conventional problems like dispersion and fragmentation will have more significant role such that an effective solution for resource management in future multicore architectures will be a necessity. What motivated us to use the clustered-mesh topology was to improve performance through sharing capability of on-chip network resources among cores. For the first time, Balfour and Dally in [36] proposed a concentration mesh topology in on-chip networks (*Cmesh*). Also Kumar in [37] compared different degrees of concentration and showed that a concentration with degree 4 will lead to more symmetry and less execution time compared to other ones. However, because of higher probability of contention, designing a high-performance on-chip network will require efficient sharing of on-chip network resources such as buffers and wire bandwidth. Moreover, limited power budget and thermal constraints do not let excessive use of resources.

Thermal design power (TDP) provides a single and constant value as power constraint and causes big performance losses in many-core systems [38]. Since the power consumption is a function of the number of simultaneously active cores, a variable and realistic upper bound for it such as thermal safe power (TSP) [39] makes the calculated power budget as optimistic as possible which eventually better addresses the issues of dark silicon phenomenon.

3. SCMesh: A SCALABLE AND HIGH BANDWIDTH NoC

As stated earlier, with the growing number of cores in a chip, power consumption is also increased. Most state-of-the-art design methodologies have not considered dark silicon issues; they try to fully utilize on-chip resources without considering power limitations. To properly address dark silicon issues, a new platform is needed that supports the following requirements:

- Achieving high performance for running applications and considering different application's communication needs.
- Managing power consumption of different components of the network and gain a higher power budget.

As the performance gain is directly related to application behavior, assigning a fixed bandwidth to each application without considering its communication requirements affects the overall performance of applications with higher communication requirements and waste the resources assigned to applications with lower communication needs. So, case-based bandwidth allocation which is tailored to the needs of an application have significant impact on the system performance. To this end, we propose a concentrated mesh network, based on *Cmesh* topology [36], called *SCmesh*, with the capability of sharing resources, in which each router is connected to four cores (cores 0–3 in Fig. 1A) forming a cluster and clusters are interconnected by a 2D-mesh network. Since *SCmesh* (like *Cmesh*) uses less routers than the conventional mesh, it provides lower messaging hop count and wiring complexity [40]. As shown in Fig. 1B, there are five ports (4 input $4w$-bit ports and one local $4w$-bit port, w being the flit width) and 20 network input buffers arranged in 5 ports (i.e., four w-bit slots at each port). On the other hand, in an equivalent conventional mesh NoC interconnecting the same number of cores, the number of routers is quadrupled and each router has five w-bit ports and five $4w$-bit input buffers (i.e., one $4w$-bit buffer at each port). Using Synopsys Design Compiler [41] with 45 nm technology file shows that the area overhead of *SCmesh* is 2.13% compared to baseline mesh network in a 64-core system. Fig. 1 shows the *SCmesh* architecture and its router structure.

Another difference of our proposed platform with a 2D-mesh topology is its fourfold link bandwidth due to 4-core connections to routers. For example, if 2D-mesh network link's bandwidth is w bps, the proposed network link's bandwidth will be $4w$ bps. This highlights the flexibility in assigning resources of each cluster. More precisely, the bandwidth which is allocated

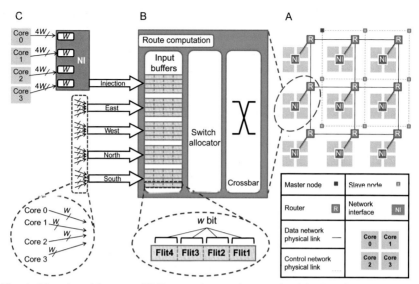

Fig. 1 *SCmesh* architecture: (A) Data and control networks. (B) *SCmesh* router micro-architecture. (C) Network interface.

to each core can vary from *1w* to *4w*, according to the core needs and conditions in each cluster. This means that each core has initially its own *w*-bps bandwidth. Regarding this fact, if each core within a cluster needs more bandwidth and other cores within the same cluster do not need their own bandwidth, their bandwidth can be assigned to the demanding cores. Accordingly, the core that has more bandwidth can forward its packets in fewer cycles and increase the total performance.

To prevent complexity issues, it is assumed that the bandwidth allocated to each core in a cluster can be:

− *w*-bps: when a core cannot use any other core's bandwidth.

− *2w*-bps: when core 0 does not require its own bandwidth and assigns its bandwidth to core 1. Therefore, core 1 can use links with *2w*-bps bandwidth, and vice versa. The same policy is used between core 2 and core 3, too. But it is not used between other pairs of cores for the sake of simplicity.

− *4w*-bps: when three cores within a cluster do not have any requests for using their bandwidth. So, the fourth core transfers its packets through the link with *4w*-bps bandwidth. For example, if core 0 wants to send a large amount of data while the others (i.e., core 1, core 2, core 3) do not require their resources, links with *4w*-bps bandwidth is assigned to core 0.

Therefore, this technique produces better results for workloads which have diversity of communication requirements.

It is worth to mention that all channels assigned to each core, exclusively belong to that core during packet forwarding.

Due to connection of four cores to a router and concurrent transferring of data through links with maximum $4w$-bps bandwidth, we need to change network interface (NI) in order to support resource sharing capability (Fig. 1C). To address this issue, each core should be able to use any other core's buffer. Therefore, we consider $4w$-bit buffer capacity and a controller for managing this buffer in each NI. The controller first checks 2-bit requests and then assigns bandwidth to each core, according to the core's request and the usage status of bandwidth by other cores. So, each core can write data into NI buffers according to its associated bandwidth (three cases mentioned above). Fig. 1A and B shows the proposed router structure with pipeline stages and NI. As shown in the router's pipeline diagram (Fig. 1B), each flit of a packet proceeds through routing computation (RC), switch allocation (SA), and switch traversal (ST) stages, respectively.

3.1 Power Consumption Analysis

Many studies have already been focused on using low-power techniques in NoCs. Nevertheless, it is important to note that in low-power systems, components are not necessarily powered off and the main purpose of such systems is power reduction. In dark silicon era, however, in addition to the existence of low-power components, turning off part of a chip is a necessity. In other words, decreasing the power of a component enables us to use a greater part of the chip.

Regarding significant contribution of interconnection network in power consumption, it is necessary to manage power consumption in different components of our proposed architecture. To this end, we focus on power reduction of buffers. This is due to the fact that the main fraction of NoC's static power is consumed in the routers [42], and largest contributors of power consumption in routers are buffers [22, 43]. Table 1 indicates static and dynamic power of different components of the *SCmesh* router.

As Table 1 illustrates, a large fraction of *SCmesh* router's power consumption is dedicated to buffers. Thus, turning off unused buffers greatly reduces power consumption. To this end, we use power gating (PG), which

Table 1 Static and Dynamic Power Decomposition of *SCmesh* Router

Power Components	Static Power (mW)	Dynamic Power (mW)
Buffers	51.97	153.56
Crossbar	7.42	54.84
Switch allocation	1.54	4.38
Route computation	0.9	6.58

is a well-known technique for saving leakage power [44]. In this method, we use two levels of granularity for power gating of the unused components, buffers and the whole router.

Although dim mode (under clock) can be used along with turned on and off modes, we only use the last two modes due to consuming a lot of static power in dim mode. This comes from the fact that by scaling technology to deep submicron processes, leakage power becomes increasingly significant compared to dynamic power [45].

3.2 Managing the Control Signals

In order to manage the power consumption of network nodes, using a control network besides data network is required. Data network is responsible for delivering data packets among PEs. Control network (routers and links represented by smaller squares and dotted lines in Fig. 1A) is used to turn on/off both buffers and routers of data network. This network should be always turned on in order to control data network. Therefore, it must be a light-weight network with low power and bandwidth. This requirement is in line with the definition of control network, in which simple routers are used for transferring small control messages.

In this network, there are slave nodes corresponding to each cluster of data network. Also, a global master node, *GM*, is aware of the on/off status of all PEs, and in case a cluster is needed to be turned on based on data network requirements, forwards a control signal to the corresponding router in control network. The communication volume for control messages is [some bits (for showing cluster addresses, it depends on network size) + 1 bit (cluster status) + 2 bits (4 cores per cluster status) + 7 bits (for 20 buffer with 4 flit)] × MD (Manhattan distance of all PEs to *GM*). Meanwhile, control network imposes 2.19% area overhead with respect to the data network in a 64-cores system, according to the synthesized results from Synopsys Design Compiler [41] at 45 nm technology.

4. STRATEGY 1: REVISITING PROCESSOR ALLOCATION

Fig. 2 shows an overview of the proposed method giving a perspective of the Dark Silicon Aware Processor Allocation (DSAPA) steps and their goals. In our scheme, a chip is divided into some sections. A *section* consists of some clusters. Each *cluster* includes four processors with dedicated memory and one router. *Region* is an area of one section including clusters that are selected by processor allocation algorithm. *Node* is a general concept referring to either cluster or section.

In fact, we assume a N-core chip is partitioned to K sections each containing M cores, so, there exist $N = K \times M$ cores in the chip. Also, a percentage of cores, say B, can be activated simultaneously in each section (dark silicon percentage). The number of required cores for each application is n_i.

Regarding our section-based approach, the first problem is to determine the section number on the chip, the number of cores per section, and initial active cores in each section. The number of powered-on cores in each section must be determined based on the power budget. Esmaeilzadeh et al. expressed dark silicon gap results in 21% and more than 50% of chips will be dark at 22 nm and 8 nm, respectively [46]. So, dark silicon percentage determines the active fraction of the chip. Initially, all sections start with almost equal dark area. If the technology constrains to power-on half of the chip (50% dark area), each section can use half of its resources at the beginning. Assuming that the number of required cores are given and fixed, to guarantee that all resources belonging to one application are placed in one section, we take a conservative approach and allocate a sufficient number of cores for applications ($Max(n_i)$). However, for applications with dynamic extra requirements, we dedicate nearby sections to minimize global intercommunications.

The second problem is how to assign needed cores to each application. We assign cores to the application based on Eq. (1). This scheme allows for using up the resources and provides desired performance through dedicating undemand power budget of one section to others. In Eq. (1), application requirements and its allocated resources are shown by n_i and $alloc_i$, respectively. If the application requires n_i cores and n_i is less than or equal to its section budget ($B \times M$), n_i cores will be assigned to it. If the workload needs more than the base budget ($B \times M$), and allocated resources to previous applications and required resources to app_i are less than the total expected budget for these sections ($B \times M \times i$), n_i core can be assigned to application

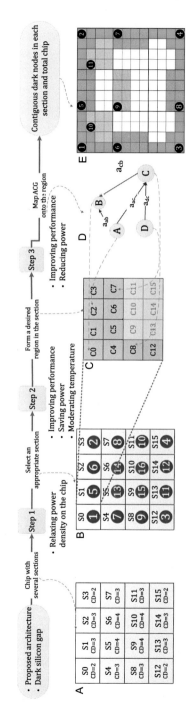

Fig. 2 Overview of the proposed processor allocation algorithm.

in order to avoid intersection communications. Otherwise, we assign up to $B \times M$ cores to the current application (app_i).

$$alloc(i) = \begin{cases} n_i & \text{if } n_i \leq B \times M \\ n_i & \text{if } n_i \geq B \times M, \sum_{j=1}^{i} alloc_j \leq B \times M \times i \\ B \times M & \text{if } n_i \geq B \times M, \sum_{j=1}^{i} alloc_j \geq B \times M \times i \end{cases} \tag{1}$$

Some terms are defined as follows:

Connectivity Degree (CD): Neighbor count of each node is the CD of the node.

Dark Degree (DD): Number of inactive adjacent nodes of a given node is its DD. The smallest value indicates the best candidate for supporting contiguous region. This value is changed during region formation process, contrary to CD.

Section ID and Cluster ID: The node address consists of two parts, the section number and the cluster number. Both are assigned by row–major indexing (e.g., first and the second clusters in section#0 are specified by (0,0) and (0,1)).

Occupied Cluster Count (OCC): Number of occupied clusters in each section.

In step 1, an appropriate order of sections among the available sections must be chosen aiming to avoid hotspots. Since the center of the chip is overheated potentially, marginal sections are selected as much as possible. In other words, Algorithm 1 produces the order of sections as an output (O_{List}) for a list of input applications (A_{List}) and chip sections (S_{List}) containing section ID, CD, OCC. S_{List} is sorted in ascending order based on connectivity degree and section ID as the first and second keys to find the proper order. After assigning one application to the chosen section, the section goes to the end of the list for better balancing sections utilization. When an application is arrived and there is no free section, the algorithm searches for a section with enough clusters to run the current application. If it cannot find an appropriate section, it waits until a suitable section becomes available.

Each section remains in the list until all its clusters are allocated. Applications release their occupied resources after termination. Fig. 2A indicates the sections of the chip and Fig. 2B shows their selection order (i.e., O_{List} elements).

ALGORITHM 1 Step1

1 **Input**: App_{List} contains unallocated applications.
2   Sec_{List} represents sections (ID, CD, OCC).
3 **Output**: $SelSec_{List}$ is an ordered list of selected sections on the chip.
4 $N_{Cluster}$: Number of clusters in each section.
5 Sort (Sec_{List}, Section CD, Section ID);
6 $i \leftarrow 0$;
7 **while** *unallocated application* **do**
8   Select *app* from App_{List};
9   **while** *app is not allocated* **do**
10   **if** *(there is space for app_i in $Sec_{List}[i]$)* **then**
11              $Sec_{List}[i].OCC + = N_{Req.Cluster}$;
12              $SelSec_{List} \leftarrow Sec_{List}[i].ID$;
13   move $Sec_{List}[i]$ to tail;
14   remove *app* from App_{List} ;
15   **if** *(there is empty section)* **then**
16                  $i \leftarrow i + 1$;
17   **else**
18                  $i \leftarrow 0$;
19   **end**
20   **end**
21   **end**
22 **end**

In the second step, our objective is to form a contiguous region of active clusters in one section (Fig. 2C) and, at the same time, trying to have the dark clusters of the whole chip closed to each other as much as possible (Fig. 2E). In fact, in this way, we benefit from contiguous processor allocation for both current and upcoming applications. This way, we must sort clusters in the given section (output of step1) based on cluster CD values at first, cluster DD values and cluster IDs as the second and third keys.

Fig. 3 illustrates the process of forming a region (Algorithm 2) in section#0 (Fig. 2C) containing 16 clusters for the current application requiring 10 clusters. In this example, the first application runs on the system and

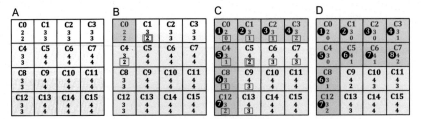

Fig. 3 Region forming in given section (section #0): (A) CD, DD, and cluster ID values for the initial state of section #0, (B) the first choice and its effect on DDs value, (C) the state of mentioned value after seven selections, and (D) the final state.

ALGORITHM 2 Step2

1 **Input**: $Clus_{List}$ represents sections (ID, CD, OCC).
2 **Output**: Req_{List} is an ordered list of selected clusters forming a region on the selected section(Step 1).
3 *Clus.status*: Active/Dark
4 Sort (*$Clus_{List}$*, Cluster CD, Cluster DD, Cluster ID);
5 **while** *application needs more clusters* **do**
6   Select *clus* from $Clus_{List}$;
7   **if** (*Clus.status* = *Dark*) **then**
8   *Clus.status* = *Active*;
9   Reg_{List} ← *clus*;
10   Update $Clus_{List}.DD$;
11   **end**
12 **end**

---

neighbor sections #1 and #4 are free. Fig. 3A, demonstrates identical values for both CDs and DDs which is the initial state. Among all clusters, only the CD value of cluster#0 is 2, so it is the first choice. Now, DD value of cluster#1 and cluster#4 are updated (Fig. 3B). Then, no cluster with CD=2 is remained and it is the turn of nodes whose CD=3. Regarding cluster ID, we pick cluster#1 and then clusters#2, 3, 4, 8, and 12, respectively. After selecting all clusters with CD=3, clusters with CD=4 will be investigated. Fig. 3C shows that the only candidate is cluster#5 with lowest DD value.

Afterward, cluster#6 and cluster#9 have the same SD. Cluster#6 is preferred due to its lower cluster ID. Finally, cluster#7 is chosen as shown in Fig. 3D.

Fig. 2E depicts the system behavior with 9 sections to run 11 applications requesting 8 clusters (allowed by the power budget) where white dots represent inactive area. The start point of the regions is represented by dark circles mentioning their selection order (which is expressed as first step). Although increasing section size leads to forming non-convex regions, the arrangement of dark nodes in DSAPA provides acceptable results for highly multiprogram multithreaded system.

The third step maps the tasks of an application onto the determined region (the output of the second step) with the aim of improving performance and power efficiency by exploiting the features of the platform. To do so, we try to minimize the communication overhead in this step through incorporating application behavior by considering application characteristic graph (ACG), similar to [47] approach. Along with it, we introduce two policies for mapping ACGs in the selected region using the capabilities of the proposed platform (i.e., clustering and sharing). The ACG is a directed graph $G(V, A)$ that defines application tasks and their execution order and relationship via vertices and arcs, respectively. A a_{ij} represents the communication from v_i to v_j, and the communication rate corresponding to this arc is denoted by $r(a_{ij})$. In [47], authors consider communication rates and Manhattan distances of resources dedicated to vertices for minimizing communication cost. We add the following policies to the process of mapping tasks onto the region. In fact, we select one of the following policies based on the application behavior, and add it to the scheme proposed in [47] as the third step of DSAPA.

4.1 Policy 1: Reducing Hop Count

Relative vertices with high communication rates (i.e., arcs with high $r(a_{ij})$) are settled in the same cluster. So, transmission latency and power consumption are decreased by avoiding forward traffic into on-chip network.

4.2 Policy 2: Sharing Resources

Locating a talkative vertex next to silent vertices improves the total performance through sharing capabilities. To comply with the infrastructure, this policy is applied for one talkative core and three inactive cores

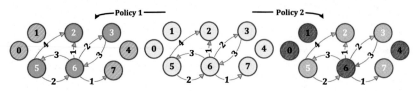

Fig. 4 Appling policies for an application: (*middle*) input ACG, outputs of (*left*) policy 1 and (*right*) policy 2.

or two talkative cores and two inactive cores on the same cluster. It performs best for jobs with diversity in communication rates.

If the application task graph shows sparsity of communicative tasks, the second policy is preferred. Otherwise, this step suggests the first policy. Our experiments reveal that the two policies provide the same average temperature (for the same application), but the second policy reduces the peak temperature by 4 K, on average.

Fig. 4 illustrates the proposed mapping policies for an application's ACG onto a region containing two clusters (eight cores). Vertex 6 has the largest communication rate to its neighbors, and thus, is the first choice. The first policy puts $v2$, $v3$, $v5$, and $v6$ in the same cluster (Fig. 4, left). On the other hand, regarding both high communication rates of $v6$, and low communication rates of $v0$, $v1$, and $v4$, the second policy suggests putting them in one cluster (Fig. 4, right).

To solve the contention, DSAPA intends to select contiguous groups of resources in suitable sections for parallel jobs. Additionally, it addresses internal contention through cooperating application behavior in mapping phase.

4.3 Overhead of Allocation Algorithm

In both, the first and second step of DSAPA algorithm (Algorithms 1 and 2), a sort operation must be done which is not a performance bottleneck. If there are m sections, Algorithm 1, for each application, performs a maximum number of m searches to find a section. Hence, for n applications, it has a polynomial complexity of $m \times n$. The second algorithm searches in a list to find a cluster (with linear complexity). Note that these algorithms execute once before running the applications. In comparison to execution time of evaluated multithreaded applications, its execution overhead is negligible (less than 1%). Additionally, since our approach does not need a dynamic profiler, it does not rely on OS to work. In [47], it is shown that UA has also negligible overhead. So, our third step which is based on UA will have almost no overhead.

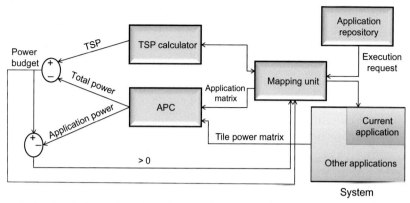

Fig. 5 Top-level abstraction of implemented system architecture.

5. STRATEGY 2: REVISITING APPLICATION MAPPING

5.1 Top-Level Abstraction of System

As shown in Fig. 5, first, the available power budget of system is computed by calculating the total power of the chip and current power budget of the system extracted from TSP library. Afterward, it is compared with the application power estimated by application power calculator (APS), according to the example provided in [16]. If the system has enough power budget left, the mapping unit allocates network resources to incoming application tasks, the application starts executing on PEs (processing elements, or cores) and the system configuration is updated. In this method, cores are allocated to new arriving applications dynamically, but when a new application's tasks are mapped onto some PEs, the mapping is fixed and cannot be changed until its execution is completed. That is an application may arrive, be mapped and executed, and exit the system, and later on again reenter the system and be mapped onto another place in the system. So, the proposed solution treats arriving applications dynamically but does not allow preempting and remapping of the applications during execution.

5.2 Algorithms for Dark Silicon Aware Mapping

The objective is to map task graph vertices onto different clusters of *SCmesh* architecture in an optimal way to minimize the total execution time and to meet power budget of the system. Therefore, we formulate this problem in a more formal way.

Definition 1. An application characterization graph, ACG (V, E), is a directed graph, where a vertex $v_i \in V$ represents a core, and an edge $e_{i, j} = (v_i, v_j) \in E$ characterizes the communication from v_i to v_j. The following properties are related to the graph:

- $w(e_{i, j})$ defines the communication weight corresponding to $e_{i, j}$, which is based on the number of packets that are delivered from v_i to v_j and can have w, $2w$, and $4w$ values.
- $ws(v_i)$ characterizes summation of $w(e_{i, j})$ in which v_i has communication with v_j, $\forall v_j \in V$.

Definition 2. Properties of a $SCmesh$ architecture with $N \times M$ clusters:

- $Cluster(I)$: Particular cluster for vertex v_i.
- $MD(Cluster(I), Cluster(J))$: Manhattan distance between clusters where vertices v_i and v_j are allocated. Distance between the PEs of a cluster is considered to be zero.
- SMC: Summation of mutual communication between the vertices mapped in a cluster, calculated by:

$$\sum\nolimits_{\forall v_i, v_j mapped\ in\ Cluster(I)} w(e_{i,j}) \tag{2}$$

If $Cluster(I) = (x_1, y_1)$ and $Cluster(J) = (x_2, y_2)$, then: $MD(Cluster(I), Cluster(J)) = |x_1 - y_1| + |x_2 - y_2|$. Task mapping problem is formulated as follows:

Given a region R and the ACG of incoming application
Find $Cluster(I)$ inside R, $\forall v_i \in V$ which:

$$\sum\nolimits_{\forall e_{i,j} \in E} MD(Cluster(I), Cluster(J)) \times w(e_{i,j}) \tag{3}$$

Definition 3. There are four registers, R_E, R_W, R_N, and R_S, used for keeping track of the cores occupying buffers in a cluster. Each register has 4 bits corresponding to east, west, north and south ports of a router, respectively. In these registers, by default, 0th, 1st, 2nd, and 3rd bits are assigned to core 0, core 1, core 2, and core 3, respectively. Each bit in each register indicates that whether input buffer in the adjacent router has been devoted to corresponding core of that bit. In more details, if each of 4 input buffers of the adjacent routers is dedicated to the packet, corresponding bit of the core that forwarded packet is set to 1 and if any of the buffers is empty, it is set to 0.

These registers are also used in resource sharing capability of a cluster. For example, in case of $2w$-bps sharing, if core 0 (core 1) requires to use the input butters of the adjacent routers dedicated to core 1 (core 0), and obviously, core 1 (core 0) does not need its own buffer, both bit 0 and bit 1 set to 1. The same applies to core 2 and core 3 in each cluster. To fluently explain mapping algorithms, we use Fig. 6 to display registers of a router.

These registers are also used in resource sharing capability of a cluster. For example, in case of $2w$-bps sharing, if core 0 (core 1) requires to use the input buffers of the adjacent routers dedicated to core 1 (core 0), and obviously, core 1 (core 0) does not need its own buffer, both bit 0 and bit 1 set to 1. The same applies to core 2 and core 3 in each cluster. To fluently explain mapping algorithms, we use Fig. 6 to display registers of a router.

In both mapping algorithms presented in this chapter, forwarding packet is done with w, $2w$, or $4w$ rates if there are one, two or four unset bits in the register corresponding to the downstream router, respectively. The unset register bits that are initial conditions for possibility of forwarding packet, are named as *USRegs*. Table 2 shows how to set *USRegs*.

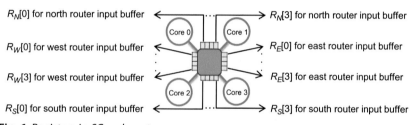

Fig. 6 Registers in *SCmesh* router.

Table 2 How to Set *USRegs* in the Router

Producer Requested Bandwidth	Location of Producer in the Cluster	USRegs Corresponding to Packet Direction			
		East	West	North	South
w	Core 0	$R_E[0]$	$R_W[0]$	$R_N[0]$	$R_S[0]$
	Core 1	$R_E[1]$	$R_W[1]$	$R_N[1]$	$R_S[1]$
	Core 2	$R_E[2]$	$R_W[2]$	$R_N[2]$	$R_S[2]$
	Core 3	$R_E[3]$	$R_W[3]$	$R_N[3]$	$R_S[3]$
$2w$	Core 0 or Core 1	$R_E[0, 1]$	$R_W[0, 1]$	$R_N[0, 1]$	$R_S[0, 1]$
	Core 2 or Core 3	$R_E[2, 3]$	$R_W[2, 3]$	$R_N[2, 3]$	$R_S[2, 3]$
$4w$	Core (0 or 1 or 2 or 3)	$R_E[0–3]$	$R_W[0–3]$	$R_N[0–3]$	$R_S[0–3]$

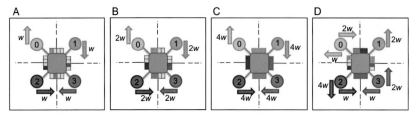

Fig. 7 Illustration of four scenarios of setting *USRegs* in a cluster.

Fig. 7 illustrates several scenarios to clarify how to set *USRegs* bits for cores in a cluster. In all of them, cores color and *USRegs* bits that are assigned to cores are matched for better understanding. Also, next hop direction of each core is displayed by arrows and requested bandwidth of each core is specified beside the arrows. In Fig. 7A, cores 0, 1, 2, and 3 want to send their packets by w-bps bandwidths to the north, south, east, and west, respectively. So, $R_N[0]$, $R_S[1]$, $R_E[2]$, and $R_W[3]$ are set to 1 for cores 0, 1, 2, and 3, respectively. Scenario (a) is repeated with $2w$-bps requested bandwidth for each core in Fig. 7B. According to Table 2, in this case cores 0, 1 and cores 2, 3 can use each other's bandwidths. Therefore $R_N[0, 1]$, $R_S[0, 1]$, $R_E[2, 3]$, and $R_W[2, 3]$ are set to 1 for cores 0, 1, 2, and 3, respectively. In Fig. 7C scenario, same requests are done with $4w$-bps bandwidths requirements. So, all *USRegs* are set to 1, because cores need the whole bandwidth for sending packets. Fig. 7D shows a mix of different bandwidth requirements. Core 0 wants to send two packets with w-bps and $2w$-bps to west and east, respectively. Thus, $R_E[0, 1]$ are set to 1 for east and $R_W[0]$ is set to 1 for west direction. Other cores take similar actions regarding their bandwidth requirements.

Definition 4. Some variables used by the proposed schemes include
- *Mlist*: The list of mapped vertices. Initially, this list is empty.
- *Vlist*: The list of all vertices.
- *SET*: A table in which the rows are sorted by $ws(v_i)$, $v_i \in V$ *list*, that contains: (a) the ordered pair of communication rate of corresponding $e_{i, j}$ and consumer v_j, $(R_{i, j}, v_j)$, that means producer v_i communicates with consumer v_j with rate R. (b) $ws(v_i)$ for every vertex
- In both algorithms, the vertex that has maximum communication demand is placed into one cluster with maximum number of *UN*, defined as:

$$UN = |USRegs| \times |NeighborsinR| \qquad (4)$$

- MV: The number of vertices that are mapped in each cluster.
- $MARK$: A flag which indicates whether there is a possibility of mapping vertices in the cluster.

In Sections 5.2.1 and 5.2.2, we describe two algorithms for task mapping in dark silicon era based on *SCmesh* architecture. It must be noted that regions considered as the input of the problem might have more cores than the number of task graph vertices. The mapping algorithm specifies combinations of cores in the floorplan to activate. Thus, there are some active and inactive cores in the system where the active cores are in the execution mode while inactive ones are in low-power mode, e.g., power-gated.

5.2.1 Dark Silicon Aware Algorithm With High Capability of Power Saving

In this algorithm, our target relies on putting vertices with high communication together. Also, in order to meet the limited power consumption at the chip level, we restrict the communications between cores in each cluster. In this case, the total weight of communication between two cores in a cluster (*SMC*) must be no more than $4w$. The main reason for this choice is related to the fact that injection channel width for each core and the total injection channel width for four cores in a cluster are equal to w bps and $4w$ bps, respectively. Algorithm 4 represents *HCPS* mapping scheme.

Fig. 9 illustrates the process of mapping the application ACG (see Fig. 8) with *SET* (see Table 3) onto a given region by *HCPS* algorithm.

As shown in Fig. 9A, firstly v_7, which has the largest $ws(v_i)$, is assigned to v_{MAX} and is mapped in the cluster with the largest UN (tagged with v_{MAX}). Next, since SMC of v_7 and v_3 (which has maximum communication with v_7) is smaller than $4w$, v_3 is mapped to the other PE in the same cluster. Now, vertices v_2, v_8, and v_6 have the largest communications to v_7 and since SMC in this cluster is equal to $4w$, we assign PEs of adjacent clusters tagged with v_x to them and set $USRegs$ for delivering messages from v_7 to v_2, v_8, and v_6. It is impossible to map v_5 and v_{10} in this step, because all register bits belonging to v_7 have been set. So, as shown in Fig. 9B, after mapping v_7, we select next producer vertex (v_2) which has the largest $ws(v_i)$. Afterward, v_2 is assigned to v_{MAX}, and v_{11} and v_5 are assigned to v_x and are mapped in a similar way. As v_7 has been mapped in previous step, we only set $USRegs$ for delivering messages from v_2 to v_7. In the next step, producer vertex is v_3, but we skip mapping vertices which have communication with it, because all register

ALGORITHM 3 HCPS

1 **while** $|Mlist| \neq |V|$ **do**

2   ▷ RT(Main Body)=C+O(V)*max{RT(Body1),RT(Body2)}=O($V(T+log\ V)$)

3   Assign $v_i \in$ *Vlist* with the largest $ws(v_i)$ to v_{MAX};

4   **if** $v_{MAX} \notin$ *Mlist* **then**

5   ▷ RT(Body1)=C+O($T+log\ V$)

6   Select a cluster with the largest UN and $SMC \leq 4w$ after mapping v_{MAX};

7   Map v_{MAX} to one of the PEs of the cluster;

8   Add v_{MAX} to *Mlist*;

9   **while** $\exists$ *entry* $(R_{MAX,j}, v_j), v_j \in V$ **do**

10   ▷ RT(Body2)=C+max{RT(Body3),RT(Body4)}=O($T+log\ V$)

11   Assign v_j with the largest $R_{MAX,j}$ to v_X;

12   **if** $\exists$ USRegs *for* v_{MAX} *in the cluster and* $v_X \notin$ *Mlist* **then**

13   ▷ RT(Body3)=C+O($log\ V$)+max{RT(Body5),RT(Body6)}=O($T+log\ V$)

14   **if** SMC *in the cluster* $\leq$ 4w *after mapping* v_X **then**

15   ▷ RT(Body5)=C

16   Map v_X to one of the PEs of the cluster;

17   Add v_X to *Mlist*;

18   **else**

19   ▷ RT(Body6)=C+O(T)

20   Select a cluster which minimizes the MD metric;

21   Map v_X to one of the PEs of the cluster;

22   Add v_X to *Mlist*;

23   Set *USRegs* according to Table 2;

24   **else if** $v_X \in$ *Mlist and it is outside of the cluster* **then**

25   ▷ RT(Body4)=C+O($log\ V$)

26   Set USRegs according to Table 2;

27   Remove entry $(R_{MAX,X}, v_X)$;

28   Remove v_{MAX} from *Vlist*;

ALGORITHM 4 HCRS

```
1  Cx(Counter for x)=0, Cy(Counter for y)=0;
2  repeat
3     Assign vi ∈ Vlist with the largest ws(vi) to vMAX;
4     if vMAX ∉ Mlist then
5                                        ▷ RT(Body1)=C+O(T+log V)
6        Select a cluster with MARK = 0 and the largest UN;
7        Map vMAX to one of the PEs of the cluster;
8        Add vMAX to Mlist;
9        MV ← MV + 1;
10       if MV = C − 1 and Cy then
11          ⋮
12       γ
13       MARK = 1;
14       Cy = Cy + 1;
15    else if MV = C and Cy = y then
16       MARK = 1;
17       Cx = Cx + 1;
18    if Cx = x and Cy = y then
19       Stop heuristic;
20 forever;
```

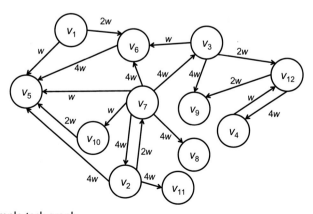

Fig. 8 Example task graph.

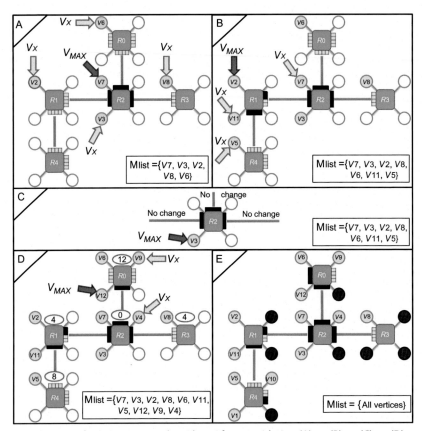

Fig. 9 Process of *HCPS* mapping algorithm. After considering (A) v_7, (B) v_2, (C) v_3, (D) v_{12} as v_{MAX}, and mapping its consumers and setting necessary registers. (E) Final result of *HCPS* mapping for the given example.

bits belonging to v_3 have been set (Fig. 9C). Then, in order to map v_{12}, we calculate *UN* for the clusters in region R. Fig. 9D shows that the cluster with *UN*=12, is selected for mapping v_{12}. We keep this process until all vertices will be mapped. Fig. 9E shows final mapping result. Black circles are representatives of turned off cores.

5.2.2 Dark Silicon Aware Algorithm With High Capability of Resource Sharing

In this method, we try to reduce the limitation of *HCPS* algorithm and take more benefits from resource sharing capability and greater bandwidth. In other words, unlike previous method where some clusters in region R may remain unused, in *HCRS* algorithm, all clusters will be used. Moreover,

Table 3 Example *SET*

Vertex v_i	$(R_{i,j}, v_j)$	ws(v_i)
v_7	$(4w, v_3), (4w, v_2), (4w, v_8),$	$18w$
	$(4w, v_6), (w, v_5), (w, v_{10})$	
v_2	$(4w, v_{11})(4w, v_5), (2w, v_7)$	$10w$
v_3	$(4w, v_9), (2w, v_{12}), (w, v_6)$	$7w$
v_{12}	$(4w, v_4), (2w, v_9)$	$6w$
v_6	$(4w, v_5)$	$4w$
v_1	$(2w, v_6), (w, v_5)$	$3w$
v_{10}	$(2w, v_5)$	$2w$
v_4	(w, v_{12})	w
v_5, v_8, v_9, v_{11}	—	0

in this algorithm, the main goal is to allow vertices with high communication volumes to place beside the vertices with low communication volumes in a cluster. Therefore, a core with more demands can often use other cores unused buffers within a router. However, in order to meet the limited power consumption budget, a certain number of cores will be mapped in each cluster. These numbers are calculated by Eq. (5) where C, x, and y represent the maximum number of usable PEs in clusters, number of clusters with C and $C - 1$ usable PE, respectively. Note that based on more usage of resource sharing capability, vertices with more $ws(v_i)$ will be allocated in clusters with less C. Therefore, this method prevents from the occurrence of hot spots in chip level. For example if we want to map an application with 20 tasks on a region with 9 clusters (36 cores), we will have Eq. (6).

$$\begin{cases} Cx + (C-1)y = number\ of\ tasks\ \ for\ C = 2\ or\ 3\ or\ 4 \\ x + y = number\ of\ clusters \end{cases} \quad (5)$$

$$\begin{cases} 3x + 2y = 20 \\ x + y = 9 \end{cases} \quad (6)$$

So, the values of C, x and y will be 3, 2, and 7, respectively. It means that 2 of 9 clusters must have 3 active PEs and 7 of 9 clusters must have 2 active PEs.

Algorithm 4 represents *HCRS* mapping scheme. Firstly, C, x, and y are calculated. Then, after each core is mapped, *MV* in each cluster counts the number of mapped cores. The algorithm is finished when the values of C_x

and C_y reach to x and y, respectively. For more efficiency, the algorithm starts mapping with the C_y parameter since due to its less coefficient, it is usually greater than C_x. By doing that, vertices with more $ws(v_i)$ will be allocated in clusters with fewer active cores. So, there is more chance to use shared resources in clusters.

We use the example of Fig. 10 to illustrate the operation of this algorithm.

The inputs of this example are the same as the previous algorithm. Initialization of all parameters are shown in Fig. 10A. As shown in Fig. 10B, we start with v_7 which has the largest $ws(v_i)$. Then, we assign it to v_{MAX} and map

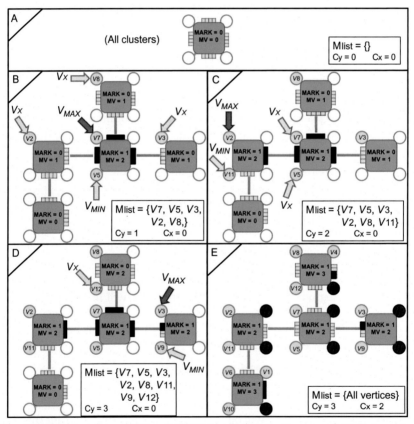

Fig. 10 (A) The status of *MV* and *MARK* in all clusters. Process of *HCRS* mapping algorithm, after considering (B) v_7, (C) v_2, (D) v_3 as v_{MAX}, and mapping low/high volume vertices and setting necessary registers. (E) Final result of *HCPS* mapping for the given example.

it to one of the PEs in the cluster with the largest UN (tagged with v_{MAX}). Then, we select v_5 which is a consumer with least $ws(v_i)$ for v_7 and map it to another PE of the cluster. Now, because MV is equal to the maximum allowed number of mapped vertices in this cluster $(C - 1)$, we mark this cluster as done $(MARK = 1)$. After that, we select v_3, v_2, and v_8 (which have maximum communications with vertex v_7) and map them to the clusters with the least MD (tagged with v_x) and set $USRegs$ for delivering messages from v_7 to them. It is not feasible to map v_6 and v_{10} at this step, because, all register bits belonging to v_7 have been set. So, as shown in Fig. 10C, after mapping v_7, we select next the producer vertex (v_2) which has the largest $ws(v_i)$, assign it to v_{MAX} and select v_{11} as v_{MIN} and map it in the same cluster. Then, we set $USRegs$ for delivering messages from v_2 to v_5 (that has been previously mapped).

Next producer vertex is v_3. Thus, we map v_9 as the consumer with least $ws(v_i)$ for v_3 in the cluster shown in Fig. 10D. Then, we assign the nearest unmarked cluster to v_{12} that has the maximum communication with vertex v_3 and set $USRegs$ for delivering messages from v_3 to v_{12}. We continue this process until all vertices will be mapped. Fig. 10E shows final mapping result where black circles represent inactive cores.

5.2.3 Overhead of the Proposed Mapping Algorithms
5.2.3.1 Time Complexity
The total runtime of our heuristics has a complexity of $O(V(T + log\ V))$. Detailed complexity analysis for calculating the runtime of each part of the pseudocodes is illustrated in Algorithms 1 and 2, where V is the number of vertices, T is the number of clusters, and C is a constant number. Since both algorithms are executed on the same data, $V\ log\ V$ is the complexity due to the sort operation in SET for $R_{i,\ j}$ and $ws(v_i)$, that works on all rows of SET concurrently and adds to RT(Main Body). Searching a vertex in $Mlist$ takes $O(log\ V)$ time, if binary search is used. Moreover, searching for a desired cluster with given parameters executes for V times in the worst case with a linear time complexity of $O(T)$.

5.2.3.2 Space complexity
Space complexity of our algorithms are of order $O(V_2 + T)$. The reason is that algorithms need at most $O(V_2)$ and $O(T)$ units for storing communication volumes information within clusters, respectively.

6. EVALUATION

6.1 Experimental Setup

We use *Sniper*, a detailed simulator for multicore systems [48], McPAT power modeling framework [49] integrated in Sniper and Synopsys Design Compiler tools [41] for performance, power and area characterization (assuming 2-cycle pipelined wormhole switched router used in [50]), respectively. The temperature profile of the baseline mesh and *SCmesh* are obtained by HotSpot tool [51]. Also, we use CACTI [52] to determine the last-level-cache (LLC) latency which is used in Sniper. Additionally, TSP library [39] is used to calculate the Thermal Safe Power. We consider an ambient temperature of 45°C and a threshold temperature of 80°C. Also, maximum power budget for the system and the power consumption of an inactive core are set to 218 and 0.14 W, respectively.

Table 4 lists a summary of the key parameters used in the evaluations. Core and cache types, sizes and area are taken from the Intel Nehalem architecture which is used in Sniper [48]. The target platform is composed of several clusters, each including 4 cores and an 8 MB shared LLC.

Table 4 Key Parameters Used in the Evaluations

	SCmesh	Mesh
Bandwidth	512 bps	128 bps
Buffer depth	16 flits per port	4 flits per port
Concentration	4	1
Frequency	2 GHz	
L1 I/D cache	32 KB/32KB, Private	
L2 cache	512 KB, Shared among (core 0, core 1) and (core 2, core 3)	
L3 cache	8192 KB, Shared among core 0, core 1, core 2, core 3	
Vdd	1.2 V	
Flit length	128 bits	
Virtual channel numbers	1	
Technology	45 nm	
Routing algorithm	XY	
Flow control	Wormhole	

Additionally, maximum Vdd and frequency have been set to 1.2 V and 2 GHz, respectively. The results are calculated for a NoC implemented in 45 nm technology.

We use 10 multithreaded applications from PARSEC [53] benchmark suite. In our experiments based on traffic traces obtained from Sniper simulator, the number of nodes per application vary from 4 to 128. Moreover, we adopt regions with sizes of 4, 9, 16, and 36 clusters. Each region is a square (i.e., connections between the routers that determine the shape of the region). We then run benchmark applications on each selected region and collect the results. Due to the lack of space, we refrain from indicating details of each application separately and only depict results for average and worst/best case of 10 applications of PARSEC [53] benchmark suite including *blackscholes, bodytrack, canneal, dedup, facesim, ferret, fluidanimate, streamcluster, swaptions, vips.*

6.1.1 Performance Analysis

For 16-thread workloads (Fig. 11A), on average, 13.1% and 12.3% performance improvements is gained by DSAPA (running on the tunable-cmesh) over FS and UA schemes, respectively. These improvements are obtained due to the features of the proposed platform and DSAPA. Hop count reduction and sharing capability are the positive points of the proposed platform in terms of performance. In addition, DSAPA avoids forwarding traffic into the NoC as much as possible via steps 2 and 3.

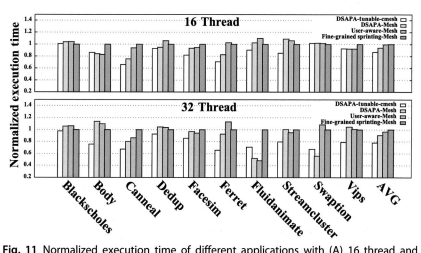

Fig. 11 Normalized execution time of different applications with (A) 16 thread and (B) 32 thread on a 64-core section.

Results of DSAPA running on the proposed platform improve the performance by 7.5% (Fig. 11A), on average, over a mesh baseline employing DSAPA. This is obtained by the two proposed methods in our platform, i.e., reducing hop count using tunable-cmesh structure and dynamically sharing the resources among idle and talkative cores.

As shown in Fig. 11B, the proposed method for 32-thread workloads improves the execution time by 21.8% and 15.5%, on average, over FS and UA schemes. A 9.8% performance gain is also obtained by DSAPA running on the proposed platform with respect to the mesh baseline employing DSAPA. Note that in Fig. 11, both 16- and 32-thread applications are mapped onto a 64-core section (with 75% and 50% dark area, respectively).

Fig. 11A and B shows that the proposed method works better for workloads with high communication rates (e.g., *canneal, ferret*) in comparison with baselines. It seems for *fluidanimate* and *swaptions*, employing the first policy (in third step) cannot provide the optimal speedup, although it saves power. We also evaluate our proposed system under multiprogram runs in a 256-core system (by randomly combining four benchmarks from PARSEC for each run) which resulted in 24.6% performance improvement.

6.1.2 Power Analysis

Fig. 12A shows that, on average, DSAPA with the proposed method of our platform (i.e., power gating undemand buffers of routers and cores completely)

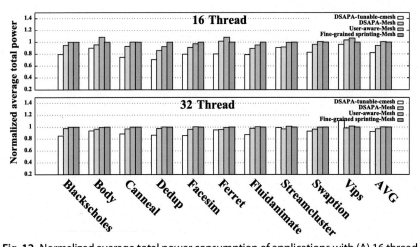

Fig. 12 Normalized average total power consumption of applications with (A) 16 thread (B) 32 thread on a 64-core section.

save the total power by 17.6% and 18.3% for 16-thread applications with respect to FS (power gating undemand routers and cores, completely) and UA (power gating mechanism is enabled for the sake of fairness), respectively. Additionally, the normalized average total power of running applications with 32 threads is shown in Fig. 12B, declaring 7.4% and 7.9% improvements over the FS and UA, respectively. Using DSAPA on the proposed platform provides 12.7% and 5.2% power reduction for 16- and 32-thread applications, on average, respectively, with respect to a mesh baseline employing DSAPA. According to Fig. 12A, *streamcluster* and *vips* consume less power, when mapped on convex regions.

We can keep NoC power consumption at the same level of FS's, on average. In fact, FS avoids turning on extra routers. However, in our scheme, although all intermediate routers outside the region are powered on to deliver messages, the used power management capability of our platform works well (by powering off undemand buffers responsible for a large percentage of router's power). Although UA is one of the best processor allocation algorithms for NoC-based CMPs, simulation results reveal that it is not a suitable allocation scheme for dark silicon designs due to allocation of noncontiguous regions (hence degrading overall performance) and formation of hotspots (leading to high power density due to mapping applications next to each other).

6.1.3 Thermal Analysis

We study the power density distribution for sections individually, and for a chip containing several sections. The average and maximum temperatures of running applications are revealed in Table 5 evaluated by HotSpot. On average, we obtain 5.7 K reduction on average which means uniform power distribution. Also, we can gain up to 11.3 K leads to avoid hot spots through our proposed platform and processor allocation algorithm.

As an example, we follow power density of *dedup* (from PARSEC suite) while it is mapped through DSAPA, user-aware, and fine-grained sprinting onto a section containing 64 cores. It is essential to mention that fine-grained sprinting has the core power gating approach, as we have. Although User-aware do not use this method for its architecture, we assume this ability to have a fair comparison. Fig. 13 demonstrates heat maps for the proposed and baseline methods. As the figure shows, we succeed in decreasing power density through putting active nodes in boundary of regions as mentioned in our algorithm. Indeed, Fig. 13A depicts the state of power

Table 5 Thermal Behavior of Applications With 16, 32 Threads on 64-Core with Different Strategies

Application	16 Thread			32 Thread		
	DSAPA[a]	UA[a]	FS[a]	DSAPA	UA	FS
Blackscholes	(329.0, 337.7)[b]	(333.8, 342.5)	(334.1, 341.9)	(337.5, 345.4)	(346.7, 355.6)	(348.2, 355.1)
Bodytrack	(331.9, 341.3)	(330.5, 354.3)	(335.6, 344.1)	(349.5, 349.6)	(349.1, 356.5)	(350.4, 357.9))
Canneal	(326.4, 332.6)	(331.2, 337.4)	(331.2, 337.2)	(334.7, 339.8)	(344.4, 350.6)	(345.2, 351.1)
Dedup	(327.1, 334.0)	(331.9, 338.9)	(333.0, 340.1)	(337.9, 343.6)	(346.9, 354.0)	(348.1, 354.2)
Facesim	(327.9, 335.8)	(332.0, 339.9)	(332.4, 339.0)	(335.6, 341.8)	(345.3, 351.8)	(346.4, 352.6)
Ferret	(327.8, 334.2)	(331.7, 338.0)	(331.3, 337.3)	(337.0, 342.4)	(345.0, 351.2)	(345.9, 351.9)
Fluidanimate	(328.7, 336.1)	(339.9, 339.9)	(333.6, 340.8)	(337.5, 334.4)	(345.6, 354.2)	(348.3, 355.0)
Streamcluster	(329.7, 337.8)	(333.2, 340.7)	(333.3, 341.2)	(341.9, 350.5)	(348.5, 356.4)	(349.7, 357.8)
Swaptions	(332.2, 341.3)	(337.1, 348.3)	(337.0, 345.9)	(338.9, 345.6)	(346.0, 352.5)	(346.8, 353.1)
Vips	(341.6, 357.6)	(345.9, 359.7)	(341.6, 354.1)	(353.2, 364.6)	(353.6, 369.7)	(354.9, 363.2)
Average	(330.2, 338.8)	(334.7, 343.9)	(334.3, 342.1)	(340.3, 345.7)	(347.1, 354.3)	(348.3, 355.1)

[a]*DSAPA*, dark silicon aware processor allocation; *FS*, fine-grained sprinting; *UA*, user-aware.
[b]Average temperature, max temperature.

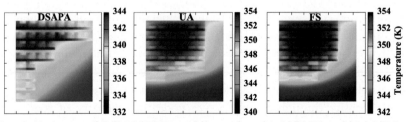

Fig. 13 Heat maps of (*left*) DSAPA in tunable-cmesh and (*middle*) UA and (*right*) FS in a baseline mesh for 32-thread *dedup* on 64-core section #0 of Fig. 3.

density for section #0, relevant to Fig. 3 of its DSAPA process. It is necessary to mention that in this figure, each heat map has its own maximum. In fact, the hotspot temperature for DSAPA (Fig. 13A) is 343.6 K, while for user-aware (Fig. 13B) and fine-grained printing (Fig. 13C) these temperatures are 354.7 and 354.2 K, respectively.

6.2 Evaluation of the Mapping Algorithm

In this section, we compare the proposed mapping algorithms that are run on *SCmesh* architecture to *PAT* mapping algorithm [17] that is run on the conventional mesh NoC [17] as the baseline architecture. *PAT* is a state of the-art mapping approach which targets dark silicon issue through patterning active cores alongside dark cores and aim to improve power budget and throughput. We consider five parameters, performance, total and NoC power consumption, power budget and temperature, in our evaluation. The first three are normalized to the results of *PAT* algorithm on the conventional mesh NoC. The main improvement of our method over the baseline comes from better mapping algorithms as well as more efficient use of the proposed cluster-based NoC.

6.2.1 Power Budget Estimation

In this section, we compare safe power constraint values of *HCPS* and *HCRS* mappings to *PAT* algorithm and the worst case mapping found by TSP_{worst}, by computing TSP. In fact, when considering TSP, each active core can have any amount of power not exceeding TSP value. TSP_{worst} provides the lowest TSP values for the worst case mapping and is safe for all mapping scenarios. In fact, we follow a conservative approach by considering the power limits provided by TSP_{worst}.

Fig. 14 shows the average power budgets of *HCPS*, *HCRS*, *PAT* and worst case mapping algorithms, which are simulated on network sizes of 4 × 4, 6 × 6, 8 × 8, and 12 × 12 (Fig. 14). Thread numbers are representatives of active cores numbers.

As seen in Fig. 14, on average, increasing the dark region in each network size results in more power budget gain. The same happens when increasing the network size. Due to higher numbers of turned off cores for a given network size and threads number in *HCPS* and *HCRS* over *PAT* algorithm, a surplus power budget can be achieved. On the other hand, the distribution of threads on cores in *HCRS* is done more scattered than *HCPS*. Therefore, heat distribution is balanced which results in lower power constraints. In fact, different power constraints for different mappings

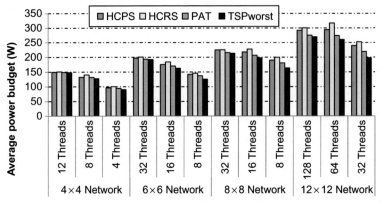

Fig. 14 Average power budget of *HCPS, HCRS, PAT,* and worst case mapping algorithms on network sizes of 4 × 4, 6 × 6, 8 × 8, and 12 × 12.

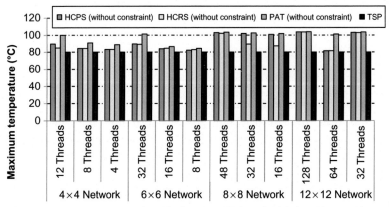

Fig. 15 Maximum temperature (in °C) of *HCPS, HCRS,* and *PAT* algorithms without any constraints and with consideration of TSP.

are consequences of heat distribution among cores. The gain is variable from 0.7% to 9.3% for *HCPS* over *PAT* and from 1.6% to 15.5% for *HCRS* over *PAT,* on average.

6.2.2 Thermal Analysis

Because of the direct correlation between power consumption and cores temperatures, managing the power upper bound results in avoiding hot-spots. So, using TSP not only makes power budget better but also provides a safe temperature at the chip level (80°C). To quantify that, the thermal profile is evaluated. Fig. 15 depicts the temperature profiles of *HCPS,* *HCRS,* and *PAT,* both with and without considering the TSP, which are simulated on network sizes of 4×4, 6×6, 8×8, and 12×12 with different thread numbers and are obtained by HotSpot tool [51].

By using TSP, on average, the temperature of the hottest spot is reduced by 15.3% and 12.1% for *HCPS* and *HCRS* algorithms, respectively. Also, without using TSP, maximum temperature of *HCPS* and *HCRS* over *PAT* is reduced by 5.5% and 8.6%, respectively. In most cases, increasing inactive cores (decreasing thread numbers) in each network size results in less maximum temperature. Because inactive cores consume lower static power.

As an example, Fig. 16 demonstrates the thermal maps of *dedup* application (from PARSEC suite) with 32 threads mapped using *HCPS*, *HCRS*, and *PAT* algorithms onto a network of 64 cores. As shown in the figure, when TSP is not used (cases A, B, and C), the maximum temperature for *HCPS* (Fig. 16A), *HCRS* (Fig. 16B), and *PAT* (Fig. 16C) are 102°C, 103°C, and 92°C, respectively. However, when TSP library is used, the maximum temperature among all cores has not exceeded 80°C (Fig. 16D–F).

6.2.3 Performance Comparison

In this section, we evaluate the performance of the proposed mapping algorithms in terms of execution time. Fig. 17A shows the average of execution times of *HCPS* and *HCRS* algorithms on *SCmesh* architecture versus *PAT* for four given region sizes.

Results reveal that the proposed strategies significantly improve the overall system performance for the power budget provided per mapping. Average execution times in each region size (16-core, 36-core, 64-core, and 144-core) for *HCPS* algorithm outperform the baseline by up to 9.9%, 17.9%, 20.9%, and 28.6%, respectively. These improvements in *HCRS* algorithm outperform the baseline by up to 14.4%, 27.9%, 29.9%, and 39.2%, respectively. In other words, the average execution time improvement of all region sizes in *HCPS* and *HCRS* algorithms are 19.3% and 27.9%, respectively. As expected, by decreasing thread numbers in each network size, average execution times decrease in all cases due to increasing the number of clusters dedicated to each application, and the capability of each core for using greater bandwidth in network. Furthermore, it is observed that *HCRS* consistently provides more performance benefits than *HCPS*. The reason is that some clusters in *HCPS* may be left unused and their routers remain off. Therefore, the average number of clusters that are used in *HCRS* is more than that of *HCPS*. This feature lets *HCRS* share more resources and increases the performance. Finally, the total average execution time in *HCRS* is 11.8% lower than that in *HCPS*.

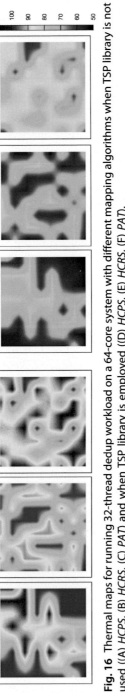

Fig. 16 Thermal maps for running 32-thread dedup workload on a 64-core system with different mapping algorithms when TSP library is not used ((A) HCPS, (B) HCRS, (C) PAT) and when TSP library is employed ((D) HCPS, (E) HCRS, (F) PAT).

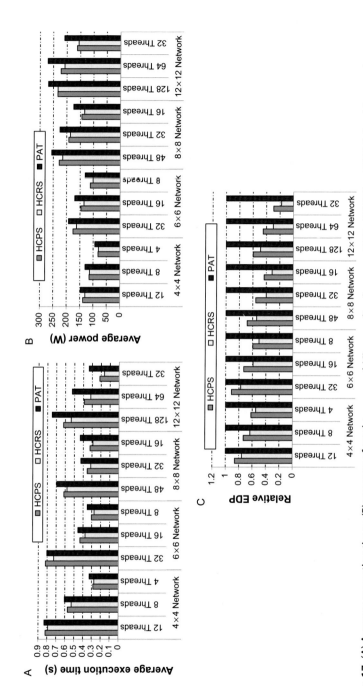

Fig. 17 (A) Average execution times, (B) average of total power, and (C) normalized average EDP of *HCPS* and *HCRS* algorithms with respect to *PAT* on network sizes of 4 × 4, 6 × 6, 8 × 8, and 12 × 12.

6.2.4 Power Consumption

In this section we evaluate the total power consumption characteristics of *HCPS* and *HCRS*, and compare those with *PAT*. Fig. 17B plots the average of total power consumption of running applications on different network sizes with different thread numbers.

The results reveal that on average, *HCPS* significantly improves the total power of the system by 11.6%, 14.9%, 18%, and 22.9% compared to the baseline for region sizes of 16, 36, 64, and 144 cores, respectively. Also, average improvements in *HCRS* for each mentioned region size are 15.6%, 21%, 24.9%, and 26.2%, respectively. Therefore, in *HCRS*, the average total power consumption for all regions is 6.5% more than the *HCPS*. The main reason for improvements of *HCPS* and *HCRS* is that more dark cores are located beside active cores, and heat effects of active cores on each other is minimized. This heat effect in *HCRS* is less due to more dispersed mapping of application tasks. Moreover, in contrast to *PAT* which uses surplus power budget to power up more cores, all inactive cores are turned off in our proposed methods and their shared resources are used to increase system performance. On the other hand, as mentioned in Sections 1 and 3, the largest contributors of the power in a router, as main consumer of NoC's static power, are buffers [22, 43]. Moreover, there is a strong relationship between static power consumption and temperature of the core. It means that higher temperature causes higher power consumption. Thus, we mostly owe our total power improvements to the turned off cores and buffers/ routers.

As a summary, according to the above results, there is a trade-off between high performance and low-power consumption for the two proposed methods depending on the desired usage.

6.2.5 Energy Efficiency

Task mapping for NoC significantly impacts the communication energy consumption of the system. Moreover, decreasing the hop counts between the communicating cores helps us to achieve the minimized energy consumption of sending data between cores. The proposed *SCmesh* platform is capable of communication bandwidth variability and two mapping algorithms that can reduce both the application execution time and the energy consumption.

Often, the goal is not to minimize energy consumption, but to find a suitable trade-off of energy savings and performance degradation. Energy-delay product (EDP) is the quantity that can be studied in this case and we compare the efficiency of the proposed methods by using this metric.

The averages of normalized EDPs of *HCPS* and *HCRS* algorithms with respect to *PAT* for four given region sizes are shown in Fig. 17C As the figure indicates, *HCRS* algorithm exhibits better average EDP compared to the baseline for all region sizes of 16, 36, 64, and 144 cores. This method locates vertices with high communication volumes beside the vertices with low communication volumes in a cluster in order to use more sharing capability and increase performance. On the contrary, *HCPS* tries to put vertices with high communication rate together and minimize communication energy. On the other hand, the small network diameter of *SCmesh* decreases packet latency with a low-energy cost. Even though the router crossbars are somewhat large, the low number of intermediate routers in most traffic patterns decrease the energy cost. Also, by coarse-grained power gating in both schemes (buffers/whole router in *HCPS* and buffers in *HCRS*) with reasonable dedicated core counts in a region, energy saving is achieved to an acceptable level. As can be observed in Fig. 17C, in networks with larger sizes and less tasks, our proposed mappings are more energy efficient than *PAT*. *SCmesh* can balance latency and energy more efficiently, resulting in a slightly lower EDP than mesh. As a result, average EDP of *SCmesh* is 43.7% less than mesh. Therefore, results show that using *SCmesh* has a great advantage in achieving energy efficiency.

7. CONCLUSIONS

In the dark silicon era, although there are large number of processing elements and huge caches available on a single chip, we cannot keep all of them ON at same time, mainly due to power budget and temperature limitations. Although there is a rich amount of research study on dark silicon-based computing (from view points of architecture, system-level or compiler-based optimizations), design and management of intercore network-on-chip is less explored. This is the scope of our research in this work. By proposing a novel and highly scalable NoC architecture, we first demonstrate the importance of NoC on scalability and computation bandwidth of a CMP with hundreds of cores. We then revisit the processor allocation strategy and application-to-core mapping algorithm to make most use of the proposed NoC architecture in order to simultaneously meet the CMP power and temperature limitation while keeping the performance of running application(s) high. The simulation results admit the efficiency of the proposed architecture—that would be about 10%–50% improvement in performance of single and multiapplication workloads.

REFERENCES

[1] N. Goulding-Hotta, et al., Greendroid: an architecture for the dark silicon age, in: 17th Asia and South Pacific Design Automation Conference (ASP-DAC), IEEE, 2012, pp. 100–105.

[2] G. Venkatesh, et al., QsCores: trading dark silicon for scalable energy efficiency with quasi-specific cores, in: Proceedings of the 44th Annual IEEE/ACM International Symposium on Microarchitecture, ACM, 2011, pp. 163–174.

[3] G. Venkatesh, et al., Conservation cores: reducing the energy of mature computations, in: ACM SIGARCH Computer Architecture News, vol. 38, ACM, 2010, pp. 205–218.

[4] T.S. Muthukaruppan, et al., Hierarchical power management for asymmetric multi-core in dark silicon era, in: Proceedings of the 50th Annual Design Automation Conference, ACM, 2013, p. 174.

[5] C.-L. Chou, et al., Energy-and performance-aware incremental mapping for networks on chip with multiple voltage levels, IEEE Trans. Comput. Aided Des. Integr. Circuits Syst. 27 (10) (2008) 1866–1879.

[6] S. Murali, G.D. Micheli, Bandwidth-constrained mapping of cores onto NoC architectures, in: Proceedings of the Conference on Design, Automation and Test in Europe, vol. 2, IEEE Computer Society, 2004, p. 20896.

[7] M.A. Bender, et al., Communication-aware processor allocation for supercomputers, in: F. Dehne, A. López-Ortiz, J.R. Sack (Eds.), Algorithms and Data Structures, Springer, 2005, pp. 169–181.

[8] A.K. Singh, et al., Communication-aware heuristics for run-time task mapping on NoC-based MPSoC platforms, J. Syst. Arch. 56 (7) (2010) 242–255.

[9] C.-L. Chou, R. Marculescu, Incremental run-time application mapping for homogeneous NoCs with multiple voltage levels, in: Proceedings of the 5th IEEE/ACM International Conference on Hardware/Software Codesign and System Synthesis, ACM, 2007, pp. 161–166.

[10] M. Mandelli, et al., Multi-task dynamic mapping onto NoC-based MPSoCs, in: Proceedings of the 24th Symposium on Integrated Circuits and Systems Design, ACM, 2011, pp. 191–196.

[11] X. Wang, et al., A power-aware mapping approach to map IP cores onto NoCs under bandwidth and latency constraints, ACM Trans. Arch. Code Optim. 7 (1) (2010) 1.

[12] D. Zhu, et al., TAPP: temperature-aware application mapping for NoC-based many-core processors, in: Proceedings of the Design, Automation & Test in Europe Conference & Exhibition, EDA Consortium, 2015, pp. 1241–1244.

[13] Y. Liu, et al., Energy and thermal aware mapping for mesh-based NoC architectures using multi-objective ant colony algorithm, in: 3rd International Conference on Computer Research and Development (ICCRD), vol. 3, IEEE, 2011, pp. 407–411.

[14] M. Moazzen, et al., CoolMap: a thermal-aware mapping algorithm for application specific networks-on-chip, in: 15th Euromicro Conference on Digital System Design (DSD), IEEE, 2012, pp. 731–734.

[15] J. Zhan, et al., NoC-Sprinting: interconnect for fine-grained sprinting in the dark silicon era, in: 51st ACM/EDAC/IEEE Design Automation Conference (DAC), IEEE, 2014, pp. 1–6.

[16] M.-H. Haghbayan, et al., Dark silicon aware power management for manycore systems under dynamic workloads, in: 32nd IEEE International Conference on Computer Design (ICCD), IEEE, 2014, pp. 509–512.

[17] A. Kanduri, et al., Dark silicon aware runtime mapping for many-core systems: a patterning approach, in: 33rd IEEE International Conference on Computer Design (ICCD), IEEE, 2015, pp. 573–580.

[18] J. Zhan, et al., DimNoC: a dim silicon approach towards power-efficient on-chip network, in: Design Automation Conference (DAC), IEEE, 2015, pp. 1–6.

[19] A. Rezaei, et al., Shift sprinting: fine-grained temperature-aware NoC-based MCSoC architecture in dark silicon age, in: Annual Design Automation Conference, ACM, 2016, p. 155.

[20] C. Li, et al., Luminoc: a power-efficient, high-performance, photonic network-on-chip for future parallel architectures, in: Proceedings of the 21st International Conference on Parallel Architectures and Compilation Techniques, ACM, 2012, pp. 421–422.

[21] M.B. Taylor, et al., The raw microprocessor: a computational fabric for software circuits and general-purpose programs, IEEE Micro 22 (2) (2002) 25–35.

[22] Y. Hoskote, et al., A 5-GHz mesh interconnect for a teraflops processor, IEEE Micro (5) (2007) 51–61.

[23] J. Howard, et al., A 48-core IA-32 processor in 45 nm CMOS using on-die message-passing and DVFS for performance and power scaling, IEEE J. Solid-State Circuits 46 (1) (2011) 173–183.

[24] D. Sanchez, G. Michelogiannakis, C. Kozyrakis, An analysis of on-chip interconnection networks for large-scale chip multiprocessors, ACM Trans. Archit. Code Optim. 7 (1) (2010) 4:1–4:28.

[25] G. Venkatesh, J. Sampson, N. Goulding, S. Garcia, V. Bryksin, J. Lugo-Martinez, S. Swanson, M.B. Taylor, Conservation cores: reducing the energy of mature computations, in: CAN, vol. 38, 2010, pp. 205–218.

[26] N. Goulding-Hotta, J. Sampson, Q. Zheng, V. Bhatt, J. Auricchio, S. Swanson, M.B. Taylor, Greendroid: an architecture for the dark silicon age, in: ASP-DAC, 2012, pp. 100–105.

[27] J. Zhan, Y. Xie, G. Sun, NoC-sprinting: interconnect for fine-grained sprinting in the dark silicon era, in: DAC, 2014, pp. 1–6.

[28] H. Bokhari, H. Javaid, M. Shafique, J. Henkel, S. Parameswaran, SuperNet: multimode interconnect architecture for many-core chips, in: DAC, 2015, p. 85.

[29] R. David, P. Bogdan, R. Marculescu, U. Ogras, Dynamic power management of voltage-frequency island partitioned networks-on-chip using Intel's single-chip cloud computer, in: NoCS, 2011, pp. 257–258.

[30] U.Y. Ogras, R. Marculescu, P. Choudhary, D. Marculescu, Voltage-frequency island partitioning for GALS-based networks-on-chip, in: DAC, 2007, pp. 110–115.

[31] H. Khdr, S. Pagani, M. Shafique, J. Henkel, Thermal constrained resource management for mixed ILP-TLP workloads in dark silicon chips, in: DAC, 2015, p. 179.

[32] V. Hanumaiah, S. Vrudhula, Energy-efficient operation of multicore processors by DVFS, task migration, and active cooling, Computers 63 (2) (2014) 349–360.

[33] V. Hanumaiah, S. Vrudhula, K.S. Chatha, Performance optimal online DVFS and task migration techniques for thermally constrained multi-core processors, IEEE Trans. Comput. Aided Des. Integr. Circ. 30 (11) (2011) 1677–1690.

[34] Y. Ge, P. Malani, Q. Qiu, Distributed task migration for thermal management in many-core systems, in: DAC, 2010, pp. 579–584.

[35] G. Liu, M. Fan, G. Quan, Neighbor-aware dynamic thermal management for multi-core platform, in: DATE, 2012, pp. 187–192.

[36] J. Balfour, W.J. Dally, Design tradeoffs for tiled CMP on-chip networks, in: Proceedings of the 20th Annual International Conference on Supercomputing, ACM, 2006, pp. 187–198.

[37] P. Kumar, et al., Exploring concentration and channel slicing in on-chip network router, in: Proceedings of the 3rd ACM/IEEE International Symposium on Networks-on-Chip, IEEE Computer Society, 2009, pp. 276–285.

[38] Intel Corporation, Intel Xeon Processor-Measuring Processor Power, Revision 1.1. In Whitepaper, Intel Corporation, 2011.

[39] S. Pagani, et al., TSP: thermal safe power: efficient power budgeting for many-core systems in dark silicon, in: Proceedings of the International Conference on Hardware/Software Codesign and System Synthesis, ACM, 2014, p. 10.

[40] W.J. Dally, Virtual-channel flow control, IEEE Trans. Parallel Distrib. Syst. 3 (2) (1992) 194–205.

[41] Synopsys, Design Compiler User Guide, 2010.

[42] L. Chen, T.M. Pinkston, Nord: node-router decoupling for effective power-gating of on-chip routers, in: Proceedings of the 45th Annual IEEE/ACM International Symposium on Microarchitecture, IEEE Computer Society, 2012, pp. 270–281.

[43] C. Fallin, et al., CHIPPER: a low-complexity bufferless deflection router, in: IEEE 17th International Symposium on High Performance Computer Architecture (HPCA), IEEE, 2011, pp. 144–155.

[44] K. Ma, X. Wang, PGCapping: exploiting power gating for power capping and core lifetime balancing in CMPs, in: Proceedings of the 21st International Conference on Parallel Architectures and Compilation Techniques, ACM, 2012, pp. 13–22.

[45] X. Chen, L.-S. Peh, Leakage power modeling and optimization in interconnection networks, in: Proceedings of the International Symposium on Low Power Electronics and Design, ACM, 2003, pp. 90–95.

[46] H. Esmaeilzadeh, E. Blem, R.S. Amant, K. Sankaralingam, D. Burger, Dark silicon and the end of multicore scaling, in: ISCA, 2011, pp. 365–376.

[47] C.-L. Chou, R. Marculescu, User-aware dynamic task allocation in networks-on-chip, in: DATE, 2008, pp. 1232–1237.

[48] T.E. Carlson, et al., Sniper: exploring the level of abstraction for scalable and accurate parallel multi-core simulation, in: Proceedings of the International Conference for High Performance Computing, Networking, Storage and Analysis, ACM, 2011, p. 52.

[49] S. Li, et al., McPAT: an integrated power, area, and timing modeling framework for multicore and manycore architectures, in: Proceedings of the 42nd Annual IEEE/ACM International Symposium on Microarchitecture, ACM, 2009, pp. 469–480.

[50] W.J. Dally, et al., Principles and Practices of Interconnection Networks, Elsevier, 2004.

[51] W. Huang, et al., HotSpot: a compact thermal modeling methodology for early-stage VLSI design, IEEE Trans. Very Large Scale Integr. VLSI Syst. 14 (5) (2006) 501–513.

[52] N. Muralimanohar, et al., CACTI 6.0: a tool to model large caches, HP Laboratories (2009) 22–31.

[53] C. Bienia, K. Li, PARSEC2.0: a new benchmark suite for chip-multiprocessors, in: Proceedings of the 5th Annual Workshop on Modeling, Benchmarking and Simulation, vol. 2011, 2009.

ABOUT THE AUTHORS

Mohaddeseh Hoveida received her B.Sc. degree from Shahed University of Tehran, Tehran, Iran, in 2010 and her M.Sc. degree from Sharif University of Technology, Tehran, Iran, in 2014, both in Computer Engineering. Her current research interests include Networks-on-Chip, Systems-on-Chip and Interconnection Networks, Multi-core and Manycore Architectures in the Dark Silicon Era.

Fatemeh Aghaaliakbari received her B.Sc. degree from Shahed University, Tehran, Iran, in 2011, and M.Sc. degree from Sharif University of Technology, Tehran, Iran, in 2014, both in Computer Engineering. She has worked as a network administrator since 2014. Her research interests include software defined networking (SDN), data center architecture, storage systems, multicore and manycore architectures, and network-on-chip (NoC).

Majid Jalili is a Ph.D. student in ECE department at the University of Texas at Austin. He received his B.Sc. degree from Shahid Bahonar University of Kerman, Kerman, Iran, in 2010, and his M.Sc. degree from Sharif University of Technology, Tehran, Iran, in 2013, where was a member of High-Performance Computing Architectures and Networks (HPCAN) Laboratory. His current research interests include memory systems, multicore and parallel computing, and heterogeneous architectures.

Ramin Bashizade is a Ph.D. student at the Computer Science department in Duke University. He received his B.Sc. from Shahed University, Tehran, Iran, in 2011, and M.Sc. degree from Sharif University of Technology, Tehran, Iran, in 2013, both in Computer Engineering. He worked as a software engineer from 2013 to 2015. His research interests include Networks-on-Chip, GPUs, and adopting emerging technologies in computer architecture.

Mohammad Arjomand received the B.Sc. degree from the Shahid Bahonar University of Kerman, Kerman, Iran, in 2006 and the M.Sc. and Ph.D. degrees from the Sharif University of Technology, Tehran, Iran, in 2008 and 2014, respectively, all in computer engineering. He held Visiting Researcher positions at Swiss Federal Institute of Technology in Lausanne, Lausanne, Switzerland (2012) and a postdoctoral at department of Computer Science an Engineering, Pennsylvania State University (2015–2017). His current research interests include multicore and manycore architectures, memory systems, storage systems, and power-aware architectures. He is currently a research engineer II and a postdoctoral fellow at school of Electrical Engineering and Computer Science, Georgia Institute of Technology.

Hamid Sarbazi-Azad received the B.Sc. degree in electrical and computer engineering from Shahid Beheshti University, Tehran, Iran, in 1992, the M.Sc. degree in computer engineering from the Sharif University of Technology, Tehran, in 1994, and the Ph.D. degree in computing science from the University of Glasgow, Glasgow, U.K., in 2002. He is currently a Professor with the Department of Computer Engineering, Sharif University of Technology, and the Head of the School of Computer Science, Institute for Research in Fundamental Sciences, Tehran. He has authored over 300 refereed conference and journal papers. His current research interests include high-performance computer/memory architectures, NoCs and system-on-chip, parallel and distributed systems, performance modeling/evaluation and storage systems. Dr. Sarbazi-Azad was a recipient of the Khwarizmi International Award in 2006, the TWAS Young Scientist Award in Engineering Sciences in 2007, and the Sharif University Distinguished Researcher Awards in 2004, 2007, 2008, 2010, and 2013. He has served as the Editor-in-Chief of the CSI Journal on Computer Science and Engineering, an Associate Editor of the IEEE TRANSACTIONS ON COMPUTERS and ACM Computing Surveys and an Editorial Board Member of the Computers and Electrical Engineering (Elsevier) journal, the International Journal of Computers and Applications and the Journal of Parallel and Distributed Computing and Networks. He is a member of the Managing Board of Computer Society of Iran.

CHAPTER THREE

Multiobjectivism in Dark Silicon Age

Amin Rezaei*, Masoud Daneshtalab[†], Hai Zhou*
*Department of Electrical Engineering and Computer Science, Northwestern University (NU), Evanston, IL, United States
[†]Division of Intelligent Future Technologies, Mälardalen University (MDH), Vasteras, Sweden

Contents

Abstract

MCSoCs, with their scalability and parallel computation power, provide an ideal implementation base for modern embedded systems. However, chip designers are facing a design challenge wherein shrinking component sizes though have improved density but started stressing energy budget. This phenomenon, that is called utilization wall, has revolutionized the semiconductor industry by shifting the main purpose of chip design from a performance-driven approach to a complex multiobjective one. The area

Advances in Computers, Volume 110
ISSN 0065-2458
https://doi.org/10.1016/bs.adcom.2018.03.012

of the chip which cannot be powered is known as dark silicon. In this chapter, we address the multiobjectivism in dark silicon age. First, we overview state-of-the-art works in a categorized manner. Second, we introduce a NoC-based MCSoC architecture, named shift sprinting, in order to increase overall reliability as well as gain high performance. Third, we explain an application mapping approach, called round rotary mapping, for HWNoC-based MCSoC in order to first balance the usage of wireless links by avoiding congestion over wireless routers and second spread temperature across the whole chip by utilizing dark silicon. Finally, we conclude the chapter by providing a future outlook of dark silicon research trend.

1. INTRODUCTION AND BACKGROUND

Embedded systems are now ubiquitous, as evidenced by smartphones, automobiles, game players, and smart appliances. Applications with ever-increasing demand of processing speed and capability for handling huge data impose formidable requirements on these systems. Many-Core System-on-Chips (MCSoCs), with their scalability and parallel computation power, provide an ideal implementation base for modern embedded systems.

However, chip designers are facing a design challenge wherein shrinking component sizes though have improved density but started stressing power budget. As a result, only some parts of the entire chip can be run at maximum permissible frequency, while the remaining parts should either be switched off (i.e., dark silicon) or run at lower frequency (i.e., dim silicon). The phenomenon of limiting performance rating, that is called utilization wall [1], has revolutionized the semiconductor industry by shifting the main purpose of chip design from a performance-driven approach to a complex multiobjective one. This multiobjectivism includes but is not limited to designing high-performance, low-power, high-reliability, and low-temperature embedded systems.

The growing interest to do more research in dark silicon field is completely beheld in recent publications. Up until now, researchers have adopted different perspectives toward the utilization wall problem that will be reviewed in this section.

1.1 The Perennial Challenge of Power Management

Power has been always a challenge in modern MCSoCs and the utilization wall problem inherently makes this challenge even more momentous.

In this regard, GreenDoid is presented for mobile applications using Conservation Cores (C-Cores) [2]. C-Cores are specialized coprocessors designed to reduce the energy athwart all the program code. They are combined with energy-efficient General-Purpose Processors (GPPs). Frequently executed pieces of the code are implemented using the C-Cores, while occasionally executed pieces are run on the GPPs. Moreover, a dark silicon architecture for servers is presented [3]. The chip area is occupied with an array of different application-specific cores. By dynamically powering up a small number of these cores, the other cores can remain dark while peak performance and power efficiency are achieved.

As an advanced approach, Computational Sprinting (CS) is proposed [4] using phase-change materials to allow cores to exceed their sustainable thermal budget for subsecond durations, providing a short but substantial computational boost. Nowadays, commercial MCSoCs are available based on Network-on-Chip (NoC) [5] communication infrastructure. It is also predicted that upcoming MCSoCs will progressively continue operating on completely new principles and novel NoC-based architectures. Thus, NoC-Sprinting (NS) is introduced [6], in which the chip selectively sprints to any intermediate stages instead of directly activating all the cores in response to short-burst computations. The mode-switching in CS lacks adaptability and only considers two states of single-core operation or all-core sprinting. However, based on the behaviors of the running applications, some in-between numbers of active cores may be sufficient for reaching the optimal performance speedup with less power consumption. On the other hand, NS lacks reliability and does not clearly consider the cooldown period required for each core after sprinting phase. By assigning sprinting periods to all or some fixed cores of the system, both CS and NS are categorized in periodical sprinting class. To address periodical sprinting problems, the concept of distributional sprinting is introduced [7], utilizing cooldown period to provide high performance for streaming applications.

Most of the papers in dark silicon field have considered heterogeneous architectures. Even the evaluation results in Ref. [8] demonstrate that building heterogeneous MCSoCs with different materials can efficiently utilize the chip-level resource and deliver the optimal balance among performance, power consumption, and cost. However, the authors in Ref. [9] have challenged the conventional wisdom that dark silicon chips must be heterogeneous in nature by building a case for homogeneous MCSoCs. The proposed solutions in this chapter are also based on homogeneous MCSoCs.

Additionally, keeping components cool is one of the most important challenges that becomes more severe by semiconductor technology scaling.

Overheating causes significant reductions in the operating life of a device. Moreover, uncertainties in reliability can lead to performance, cost, and time-to-market penalties [10]. Thus, challenges and opportunities of the emergence of dark silicon in the context of thermal management and reliability are discussed in Ref. [11]. Moreover, in Ref. [12] a new power budget concept, called Thermal Safe Power (TSP), is presented which results in a higher total system performance, while the maximum temperature among all cores remains below the threshold level that triggers dynamic thermal management.

1.2 The Essence of Efficient Application Mapping Schemes

Since the cores asymmetrically degrade in dark silicon age, the application mapping procedure is disrupted. Thus, traditional approaches can no longer satisfy the complex multiobjective goals of the system.

In Ref. [13] a sustainability control system is designed that monitors aging and passes core utilization guidelines to a Central Manager (CM) responsible for application mapping. The mapping procedure follows two main objectives: First, sustaining the benefits of differential reliability and then, maximizing energy efficiency. Moreover, since switching on all the clusters in a cluster-based MCSoC may violate power budget, an application arrival rate runtime scheduler is proposed [14] to minimize mean service time within a power budget. Applications can be migrated between clusters and typically, one cluster is active at any point, while the other clusters remain dark.

In Ref. [15] a thermal-aware application mapping policy is proposed to determine the locations of active and dark cores for each application on the chip, such that the potential temperature increase is reduced as much as possible. In Ref. [16] an efficient heuristic is employed to jointly optimize the dark silicon pattern and application mapping, which is enabled by a light-weight temperature prediction mechanism that provides an estimate of the chip temperature profile resulting from a candidate solution. In Ref. [17] a runtime application mapping approach is presented that dynamically adapts the degree of parallelism to minimize average application service times and energy.

Despite the fact that a NoC-based architecture has many advantages, its multihop nature has negative impact on both latency and power consumption parameters especially when the network size increases. Therefore, alternative technologies such as Hybrid Wireless NoC (HWNoC) have introduced [18]. Also, a temperature- and congestion-aware application mapping algorithm is presented [19] targeted at tackling two critical concerns in emerging

HWNoC-based MCSoCs: Alleviation of the severe congestion on Wireless Routers (i.e., the routers equipped with wireless transceivers, WRs) and prevention of persistent hot spots in the network.

1.3 The Emergence of Novel Perspectives

Another way to tackle the utilization wall problem is to change the traditional thinking by proposing new computing schemes, new computation components, or even new electronic devices. However, in most of the times, changing everything from basis is not an electronic factory cup of tea. Thus, the proposed solutions should be fully or partially compatible with existing technologies.

In Ref. [20] the authors have proposed to use approximate computing in programming level. However, the runtime system should be configured in a way that the approximation does not produce unacceptable outputs. In Ref. [21] the authors have identified new research opportunities in optimization of interconnects in the scenario of accelerators. In Ref. [22] the authors have used a combination of steep slope and CMOS devices in the design of MCSoCs. They have adopted the most promising steep-slope device candidate, interband Tunnel Field-Effect Transistors (TFETs), and evaluate a CMOS-TFET MCSoC.

2. SHIFT SPRINTING: RELIABLE TEMPERATURE-AWARE NoC-BASED MCSoC ARCHITECTURE IN DARK SILICON AGE

In recent years, tile-based architectures with NoC as communication framework are emerging as an attractive alternative to bus-based systems. Nowadays, users are using their smartphones not only for daily communication but also increasingly for media streaming. Thus, the Quality of Service (QoS) of real-time streaming applications will become more and more important for next generations of mobile devices. Furthermore, reliability issues are highly challengeable in the future mobile devices due to the limited cooling options. In short, designing reliable mobile devices to provide QoS-aware real-time streaming for the users is crucially important in dark silicon age. In this section, a fine-grained NoC-based MCSoC architecture, named Shift Sprinting (SS) [7], is introduced in order to reliably utilize dark silicon under the power budget constraint.

2.1 Preliminaries and Motivations

As a preliminary study, a 2×2 NoC-based SoC is simulated under uniform streaming traffic to show the necessity of high-performance and high-reliability demands of real-time streaming applications. By an optimistic assumption each sprinting period is considered to be equal to the cooldown period. Moreover, each core in sprinting status is supposed to gain performance by a factor of four and to lose life span by a factor of two compared to the core in nominal status. Maximum traffic injection rate and the average lifetime of the system running in idle status are called λ_{full} and γ_{idle}, respectively.

2.1.1 High-Performance Demands

It is shown in Refs. [4,6] that for serving short-burst computations, interleaving the status of the cores between nominal and sprinting along with cooldown intervals can be beneficial. Since in sprinting status the core is operating on higher than the Thermal Design Power (TDP) constraint, phase change of the core internal materials can be used to tolerate such situation. It is assumed that the core temperature stays constant for a specified time during the melting phase of the materials.

However, applying short-burst computations is not enough to fully support real-time streaming applications (e.g., watching a live football match) because these applications require continues high-computation demands (i.e., frequent sprinting periods) in order to provide QoS to the users. Also, as shown in Fig. 1A, applying periodical sprinting to a fixed set of cores does not solve the problem neither since it still requires cooldown intervals. On the other hand, as proposed in Fig. 1B, migrating the running application to dark cores utilizes cooldown periods and provides appropriate QoS to the users.

Fig. 2 shows the average network latency for both periodical sprinting and distributional sprinting in a 2×2 NoC-based SoC. By increasing the traffic injection rate, the weakness of periodical sprinting over distributional sprinting is fully observed.

2.1.2 High-Reliability Demands

Systems designed for dark silicon allow periodical sprinting to some specific cores while let others stay at dark. This greatly increases the permanent failure probability of those highly active cores that may lead to system failure. Hence, distributing the sprinting periods through the whole system can reduce the core malfunction possibility. Fig. 3 demonstrates the failure

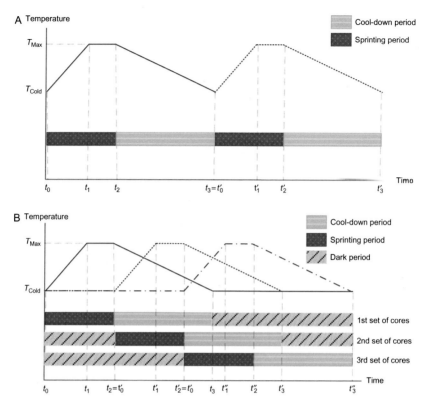

Fig. 1 High-performance demands (A) periodical sprinting and (B) distributional sprinting.

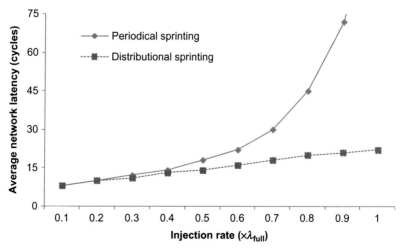

Fig. 2 Average network latency for 2 × 2 NoC-based SoC with periodical sprinting and distributional sprinting.

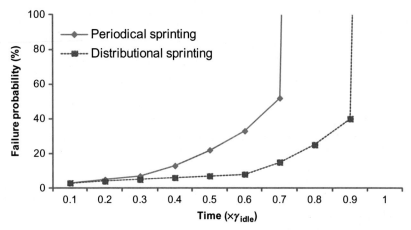

Fig. 3 Failure probability for 2 × 2 NoC-based MCSoC with periodical sprinting and distributional sprinting.

probability of the system over time in both periodical sprinting and distributional sprinting in a 2×2 NoC-based SoC under the injection rate of $0.5\ \lambda_{full}$. It is supposed that failure of a single core leads to the whole system failure. As it can be seen the failure probability of periodical sprinting is almost as $1.5 \times$ as likely as distributional sprinting.

2.2 SS Architecture

By employing the concept of distributional sprinting, both high–performance and high–reliability demands of NoC-based MCSoCs can be fulfilled in dark silicon age.

2.2.1 Core Behavior Model

The core behavior model of SS is depicted in Fig. 4. Each core has four main states including dark, idle, active, and malfunction.

Dark: In the future generations of MCSoCs, the common state of the core is dark. In this state the core is power-gated.

Idle: After waking up the core, it goes to idle state (i.e., the core is powered on but still no application is assigned to it). Moreover, after the application departure, the core goes to warm status until it cools down and reaches the cold status. Then, it can go back to dark state.

Active: When an application arrives to the core, it goes to active state. In active state the status is interchangeable between nominal and sprinting. In nominal status, the core is operating under the TDP constraint. On the

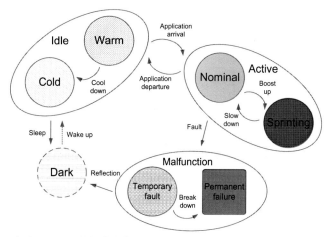

Fig. 4 Core behavior model of shift sprinting.

other hand, in sprinting status, the core is operating on higher than the TDP for a temporary period in order to speed up the process.

Malfunction: In the case of fault happening, the core state is changed to malfunction. If the fault is temporary, after resolving the problem, the core can go back into the normal cycle. Otherwise, it comes to permanent failure.

2.2.2 System Topology

Rather than abandon the benefits of transistor density scaling, some cores are transiently allowed to operate on higher than the TDP. The mobile platform trend shows that the future MCSoCs with the same die area as current mobile chips will have enough dark cores (i.e., on average 52% [1]) to support additional cores during sprinting. The topology of SS is based on the 2D mesh NoC. As an example we have eight different phases (i.e., four sprinting and four nominal) in SS shown in Fig. 5. The shift can happen between different phases whenever necessary. In nominal phases, a single core is operating under the TDP constraint. On the contrary, in sprinting phases thermal capacitance of chosen cores is increased over short timescales in order to boost up the process (i.e., the sprinted cores operate on higher than the TDP). Based on each application characteristics, the number of sprinted cores required to provide maximal performance speedup varies.

In most of the available dynamic application mapping techniques in the literature, one core is already dedicated to the CM. In SS, CM is also used to globally control application migration process. In order to speed up the application migration process, in addition to the CM, there are three

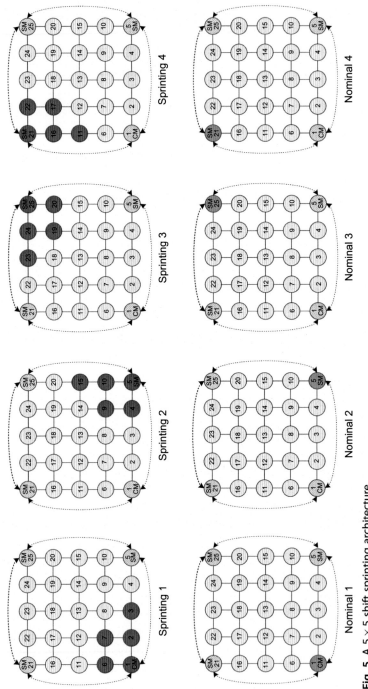

Fig. 5 A 5 × 5 shift sprinting architecture.

Sub-Managers (SMs) that are responsible for collecting information from other cores. Each core sends and receives information from its nearest manager. For controlling the application migration process efficiently, the managers have created a ring topology network. As it can be seen from Fig. 5 the CM is resided in one corner (i.e., core #1) and three SMs are resided in other corners (i.e., core #5, core #21, and core #25); they formed a ring network all together. This network can be characterized either by wireless or virtual ring network.

Wireless Ring Network (WRN): In WRN, by applying long-range, high-bandwidth, and low-power wireless links [23], a real ring network is formed on top of the 2D mesh NoC. In this case, only the manager routers are required to be equipped with wireless communication capabilities.

Virtual Ring Network (VRN): In VRN, physical channel has a virtual channel that can be dynamically configured for low-latency and low-power connection with a high priority [24]. However, all the NoC routers have to be equipped with extra virtual channels to support these connections.

2.2.3 Application Migration Scheme

If the temperature reaches a certain threshold, the application migration (i.e., shifting between different phases) will start. One upper-bound threshold for each core and a pair of upper-bound and lower-bound thresholds for the whole system are defined. SS state diagram is shown in Fig. 6.

Maximum Core Temperature (MCT): When the temperature of each core reaches the MCT, the shift happens from the current sprinting to the next sprinting phase. MCT is a static threshold relies on the core materials and is defined based on $T_{\max}$ values in Fig. 1. MCT threshold is responsible for the inner loop of Fig. 6 (i.e., shifting between different sprinting phases).

Maximum and Average Overall Temperature (MOT and AOT): When the overall temperature of the system reaches the MOT, the shift happens from the current sprinting to the next nominal phase. In reverse, when the overall temperature of the system goes back to the AOT, the shift happens from the current nominal to the sprinting phase. Both MOT and AOT are dynamic thresholds that depend highly on application behaviors. Note that obtaining the optimal values for these dynamic thresholds is beyond the scope of this chapter. MOT and AOT are managed in the outer loop of Fig. 6 (i.e., shifting between sprinting and nominal phases).

In dark silicon age, the required free cores are guaranteed to be available for application migration. With this scheme, instead of waiting for the cores to cooldown after each sprinting phase, the intervals between cooldown periods of the cores are utilized by migrating applications to dark cores

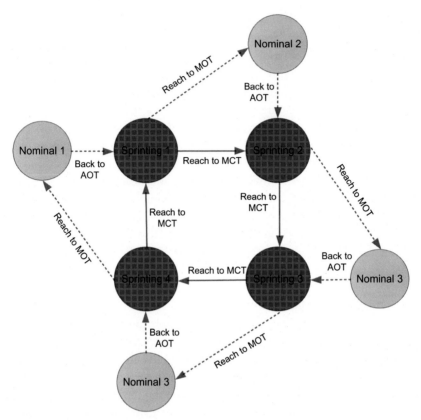

Fig. 6 Shift sprinting state diagram.

and maximizing the sprinting timelines. From evaluation perspective, SS state diagram has an inherent reliability consideration. By distributing the sprinting periods across the whole system, SS spreads the gradual aging process to all the cores. On the other hand, the more the system stays in the inner loop, the better the overall performance is.

2.2.4 Controlling Mechanism

A cost function is needed for the CM in order to determine the best destination core for application migration. When the threshold in one sprinting phase reaches the MCT, the running application will migrate to the next sprinting phase. The destination cores in the next phase are chosen based on the least current temperature order. The system stays in this loop (i.e., inner loop of Fig. 6) until it reaches the MOT threshold. For MOT threshold, all the running applications will migrate to the next phase manager until the overall temperature of the system goes back to AOT. In this case, the running

MCT: Maximum core temperature
MOT: Maximum overall temperature
AOT: Average overall temperature
C: Set of all the cores
j: Phase of the system (i.e., phases 1 to 4)
D_j: Set of the dark cores in j_{th} phase
m_j: Manager core in j_{th} phase
t_i: Temperature of the core $c_i \in C$
T: Overall temperature of the system

Set the value of MCT
Initiate phase j
while *true* **do**
    Set the value of MOT and AOT
    if *j is in sprinting phase* **then**
        if $T < MOT$ **then**
            for $\forall c_i \in j$ **do**
                if $t_i > MCT$ **then**
                    Choose $c_d \in D_{j+1}$ with $\min[t_d]$
                    Migrate the running application from c_i to c_d
            Start sprinting phase in j+1
        else
            for $\forall c_i \in j$ **do**
                Migrate the running application from c_j to m_{j+1}
            Start nominal phase in j+1
    else
        if $T < AOT$ **then**
            while $\exists$ *running application in m_j* **do**
                Choose multiple $c_{ds} \in D_j$ with $\min[t_d]$s
                Migrate the running application from m_j to the chosen c_{ds}
            Start sprinting phase in j
    Set the next phase as j

Fig. 7 Shift sprinting controlling algorithm.

applications in the manager spread through the dark cores of that phase based on the optimal number of sprinted cores required for each application. Then, the sprinting phase is started again. Fig. 7 shows SS controlling algorithm. It is assumed that only one application can be executed in a sprinted core at each time and the application itself knows the optimal number of required sprinted cores to provide maximal performance speedup.

2.3 Methodology

Experiments are performed on a cycle-accurate many-core platform implemented in SystemC. A pruned version of an open source simulator for mesh-based NoCs, called Noxim [25], is utilized as its communication architecture. For power and temperature simulations, power and thermal models taken from Refs. [26,27] are integrated as libraries into the simulator. By an optimistic assumption each sprinting period is considered to be equal to the cooldown period. Moreover, each core in sprinting status is supposed to gain performance by a factor of 4 and to lose life span by a factor of 2 compared to the core in nominal status. Maximum traffic injection rate and the average lifetime of the system running in idle status are called λ_{full} and γ_{idle}, respectively. Some multithreaded applications from the PARSEC [28] benchmark suite are used in the experiments. Three different network sizes of 16 (no dark silicon), 36 (55% dark silicon), and 64 (75% dark silicon) cores are considered in the simulations. Comparisons are also made between SS and two state-of-the-arts architectures: CS [4] and NS [6].

2.3.1 Power Model
Power consumption of each core is modeled as two major parameters, i.e., dynamic power due to transistor switching and static power due to leakage. Therefore, total power for each core is given by:

$$P_{total} = P_{dynamic} + P_{leakage} \tag{1}$$

Generally, dynamic power is given by:

$$P_{dynamic} = \alpha \, C_e \, V_{dd}^2 f \tag{2}$$

where α is the activity factor, C_e is the effective capacitance, V_{dd} is the supply voltage, and f is the running frequency of the core. Assuming both the activity factor and the effective capacitance as constants, dynamic power changes quadratically to supply voltage and linearly to frequency. According to Ref. [29], the static power of a system is given by:

$$P_{leakage} = V_{dd} \, N_{tr} \, k_d \, I_s \tag{3}$$

where N_{tr} is the number of transistors, k_d is a device-specific constant, and I_s is the normalized static current for each transistor that is proportional to the leakage current of a single transistor.

2.3.2 Thermal Model
For the thermal model, a well-known RC thermal network, defined in Ref. [27], is adopted in the simulator, which considers the duality between thermal and electrical circuits. According to this model, the steady-state temperatures of the cores can be computed as follows:

$$T^C = BP^C + H_1 + H_2 \tag{4}$$

where T^C is the steady-state temperatures of all the cores on the chip. Matrix B contains the amount of the heat contribution of all the cores. Thus, heat transfer among the cores is considered. Column vector H_1 contains the heat contribution of the other thermal nodes to the cores, while H_2 contains the heat contribution of the ambient temperature to the cores. Assuming only one application can be executed in a sprinted core, with respect to the mapping of applications, a binary matrix $L=[L_{i,\,j}]$ is defined. If application a_j is mapped to core c_i, $[L_{i,\,j}]=1$; otherwise, $[L_{i,\,j}]=0$.

By involving the power vector of the applications and the mapping matrix in Eq. (4), we get the following equation:

$$T^C = BLP^A + H \tag{5}$$

where P^A is a column vector containing the power consumptions of all the applications and H is the sum of H_1 and H_2. Furthermore, Eq. (6) expresses the direct relation between application power consumption and the steady-state temperature of any core c_i:

$$T^{c_i} = \sum_{j=1}^{k} b_i l_j p_j + H \tag{6}$$

where b_i is the row i from matrix B which corresponds to core c_i, l_j is the column j of matrix L which corresponds to application a_j, and p_j is the power consumption of application a_j. Hence, the thermal model calculates the steady-state temperature for any core of the system given the mapping matrix of the applications.

2.4 Experimental Results

2.4.1 Performance Evaluation

Fig. 8 shows the normalized execution time of different workloads in SS. In comparison with 16-core NoC (no dark silicon), the average performance improvement of 64-core NoC (75% dark silicon) is 36% and that of 36-core

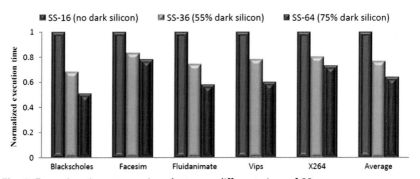

Fig. 8 Execution time comparison between different sizes of SS.

NoC (55% dark silicon) is 24%. As a result, even with increasing of dark silicon area in upcoming MCSoCs, the performance of SS will still improve. This happens because SS utilizes cooldown periods by activating dark cores.

Fig. 9 demonstrates the normalized executing time of different workloads with different architectures for 64-core NoC (75% dark silicon). The results show that SS considerably reduces the execution time compared to other approaches. It achieves 55% and 25% average performance improvement compared to CS and NS, respectively. The performance gain is because of the higher overall sprinting periods provided by utilizing cooldown periods. As it can be seen from Fig. 9, all the architectures performed quite the same under "Blackscholes" workload that is a nonstreaming financial analysis application. This is because "Blackscholes" achieves the optimal performance speedup in CS and hence leave no space for power gating in NS and neither power gating nor application migration in SS.

In addition, Fig. 10 shows the normalized network latency of different workloads with different architectures for 64-core NoC (75% dark silicon)

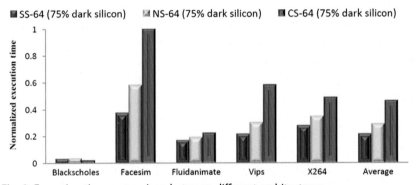

Fig. 9 Execution time comparison between different architectures.

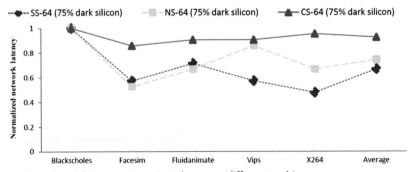

Fig. 10 Network latency comparison between different architectures.

under the injection rate of $0.75 \lambda_{full}$. It can be seen that SS reduces the communication latency for all the applications (28% in average) in comparison with CS and for most of the applications (11% in average) in comparison with NS. SS performs quite well in the media-processing applications (e.g., Vips and X264). On the other hand, performance degradation of SS in the animation workloads (e.g., facesim and fluidanimate) compared with NS is due to application migration overhead. Therefore, still more attempts are required to minimize migration overhead by finding optimal thresholds and making low-overhead migration mechanisms. Applying self-aware mechanisms based on application behavior prediction [30] may decrease the migration overhead.

2.4.2 Power Consumption Measurement

Fig. 11 displays the normalized network power consumption of different workloads with different architectures for 64-core NoC (75% dark silicon) under the injection rate of $0.75 \lambda_{full}$. On average, SS saves 58% power compared to CS while consuming almost the same power as NS. This is because of the fact that dark cores (i.e., 75% of the cores) are power-gated in both SS and NS.

As another evaluation parameter, Fig. 12 depicts the normalized Energy Delay Product (EDP) of different workloads with different architectures for 64-core NoC (75% dark silicon). It can be seen that even with the overhead of application migration approach, the average EDP of SS in the media-processing applications (e.g., Vips and X264) is less than the other architectures that makes it a promising architecture for future mobile devices.

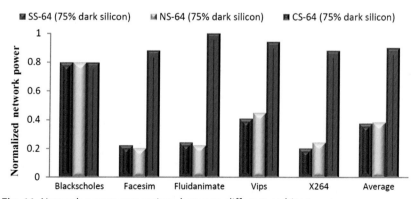

Fig. 11 Network power comparison between different architectures.

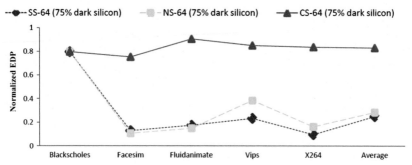

Fig. 12 EDP comparison between different architectures.

2.4.3 Thermal Analysis

Fig. 13 demonstrates the thermal analysis of SS under X264 workload for 36-core NoC (55% dark silicon) after one, two, and three consecutive sprinting. Since SS uses simultaneous techniques of core sprinting, application migration, and power gating to distribute the heat across the chip, it can efficiently avoid hot spots in the system. The peak temperature is 322.8 K and the average temperature of the system is 298.9, 304.3, and 312.5 K after one, two, and three consecutive sprinting, respectively.

Furthermore, Fig. 14 displays the thermal analysis of different architectures under X264 workload for 36-core NoC (55% dark silicon) after four consecutive sprinting. As shown in Fig. 14A, CS results in a hot spot in the center of the chip. Moreover, since thermal-aware floor planning of NS tries to physically separate logical connected cores, heat is distributed to the corners of the chip as depicted in Fig. 14B. Such floor-planning proposal has three disadvantages: First, it requires additional overheads at design stage; second, it is highly application specific and is not suitable for dynamic workloads; third, it leads to performance degradation due to long-distance communications between physically separated cores.

On the other hand, as shown in Fig. 14C, SS outperforms the other two architectures to efficiently distribute the heat across the chip. First, it does not require any temperature-aware floor planning; second, it does not rely on specific running applications to avoid hot spots; third, there is no need to change the physical positions of the cores.

2.4.4 Reliability Assessment

The central hot spot in CS and the angular hot spots in NS greatly increase the failure probability (both temporary and permanent) of those highly active cores due to frequent phase change of the core internal materials.

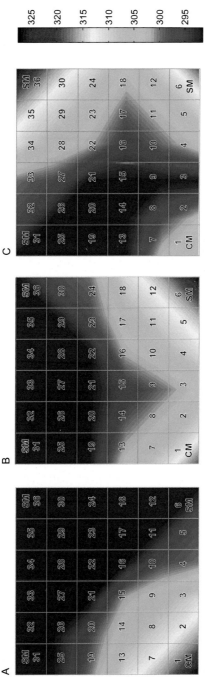

Fig. 13 Thermal distribution in SS-36 (35% dark silicon) under X264 workload after (A) one, (B) two, and (C) three consecutive sprinting.

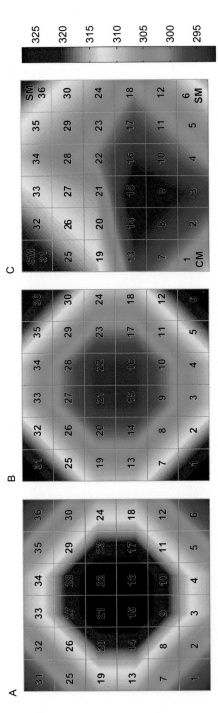

Fig. 14 Thermal distribution comparison between different architectures under X264 workload (A) CC-36, (B) NC-36, and (C) SS-36 (55% dark silicon).

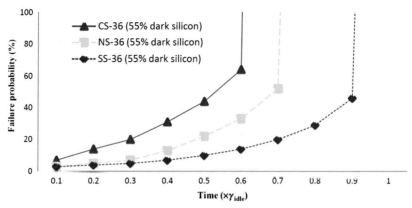

Fig. 15 Reliability comparison between different architectures.

Fig. 15 demonstrates the failure probability of different architectures over time under X264 workload for 36-core NoC (55% dark silicon) under the injection rate of $0.5 \lambda_{full}$. It is assumed that failure of a single core leads to the whole system failure. It can be seen that fair core unitization in SS not only efficiently distributes the heat across the chip but also decreases the possibility of failure of each core which ultimately increases the reliability of the system. In other words, in SS the cores are aging almost evenly. On the contrary, there are some central aged cores in CS as well as some cornered aged ones in NS through time, while the others are still young. The aged cores are increasingly subjected to failure than the young ones. This fact makes the requirement of fault-tolerant mechanisms inevitable in CS and NS.

2.5 Summary

Of all the challenges the mobile device industry faces, keeping components cool is the most important, since overheating causes significant reductions in the operating life of a device and leads to device failure. Moreover, QoS of real-time streaming applications will become more and more important for future generations of mobile devices. On the other hand, due to the dark silicon problem, the threshold voltage cannot be scaled without exponentially increasing leakage, and as a result, the operating voltage should be kept roughly constant.

Therefore, in this section, a fine-grained NoC-based MCSoC architecture, called SS, along with core behavior model, system topology, application migration scheme, and controlling mechanism was introduced in order to handle high-performance QoS-aware mobile demands by reliably utilizing

	Computational sprinting	NoC sprinting	Shift sprinting
Performance	High performance in short-burst computation demands	High performance in short-burst computation demands	High performance in media streaming demands
Power consumption	High-power consumption	Low-power consumption	Low-power consumption
Energy delay product	High EDP	Low EDP	Low EDP
Temperature	Central hot spot	Angular hot spots	No hot spot
Reliability	Low reliability	Average reliability	High reliability
Challenges	On-chip regulators	On-chip regulators, application dependency, floor-planning overhead	On-chip regulators, application migration overhead

Fig. 16 Comparison summary.

dark silicon. Simulation results reported meaningful gain in performance, temperature, and reliability of the system compared to state-of-the-art works. Fig. 16 represents a comparison summary between SS, CS, and NS. As it can be seen from Fig. 16, SS is a promising architecture for future MCSoC mobile devices capable of providing QoS media streaming demands.

3. ROUND ROTARY MAPPING: TEMPERATURE- AND CONGESTION-AWARE APPLICATION MAPPING APPROACH FOR WIRELESS NoC IN DARK SILICON AGE

HWNoC provides high-bandwidth, low-latency, and flexible topology configurations, making this emerging technology a scalable communication fabric for future MCSoCs. However, high energy costs of the WRs in comparison with the conventional routers not only limits the integration of WRs on a single chip but also introduces a direct confrontation with the utilization wall. On the other hand, by employing limited number of WRs, they are more vulnerable to congestion since far apart traffics intend to utilize wireless express links which result in high wireless channel competitions. Moreover, as it is mentioned in previous section, keeping future MCSoCs—with hundreds of embedded cores—cool has a high priority. Thus, in this section, we introduce a temperature- and congestion-aware application mapping algorithm, named Round Rotary Mapping (RRM) [19], targeted at tackling two critical concerns in HWNoC-based MCSoCs in dark silicon age: Alleviation of the severe congestion on WRs and prevention of persistent hot spots in the network.

3.1 Preliminaries and Motivations

A 4×4 HWNoC-based SoC (Fig. 17A) with two WRs is simulated using random task mapping (Fig. 17B) and congestion-aware dynamic task mapping (Fig. 17C) presented in Ref. [31] to show the necessity of both congestion avoidance and hot spot prevention. Several applications each with two to five tasks are randomly generated. Each application is considered to have 75% intracommunication among its tasks and 25% intercommunication with the tasks of other applications. Applications are scheduled based on the First-Come-First-Serve (FCFS) policy and the maximum traffic injection rate is λ_{full}. An allocation request for the scheduled application is sent to the CM of the system. CM keeps track of the free cores to map the new tasks. Moreover, each core sends its status to the CM periodically.

3.1.1 Congestion

Fig. 18 shows the average network latency comparison between random and dynamic task mapping schemes in 4×4 HWNoC-based SoC. In random task mapping, the traffic is not evenly distributed in the network, resulting in high congestion surrounding WRs which degrades the network performance significantly. On the other hand, in dynamic task mapping, the traffic is more balanced globally, resulting in great average latency reduction.

3.1.2 Hot Spot

The thermal analysis of 4×4 HWNoC-based SoC with random and dynamic task mapping in $0.5 \lambda_{full}$ is depicted in Fig. 19. In random task mapping, the hot spot regions are observed around the WRs since most of the traffics are moving toward them. With dynamic task mapping, although the congestion around WRs is decreased, the hot spot problem is getting worse around the CM because the applications are contiguously mapped as close as possible to the CM.

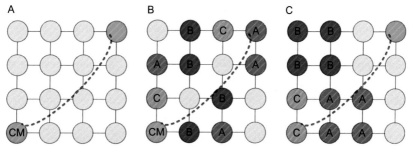

Fig. 17 (A) 4×4 HWNoC-based SoC with two WRs using (B) random task mapping and (C) dynamic task mapping.

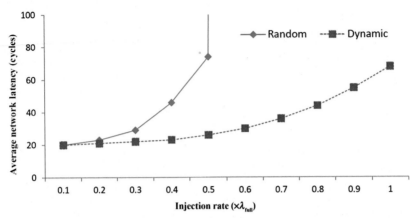

Fig. 18 Average network latency in 4 × 4 HWNoC-based SoC with random and dynamic task mapping.

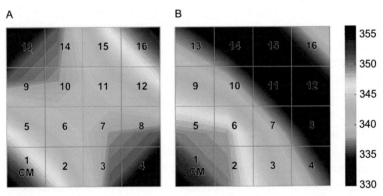

Fig. 19 Thermal distribution for 4 × 4 HWNoC-based SoC (A) random task mapping and (B) dynamic task mapping.

3.2 RRM Algorithm

Based on the above discussions, a reliable mapping is essential for HWNoC-based MCSoC that not only improves network performance by reducing severe congestion around WRs, but rather achieves energy efficiency by preventing hot spots in the network. In other words, we may combine the performance gain obtained by dynamic task mapping shown in Fig. 18 with a temperature-aware method to avoid hot spots depicted in Fig. 19.

3.2.1 System Configuration

Without loss of generality, a 2D mesh HWNoC is virtually divided into several regions where the number of regions equals to the number of available WRs.

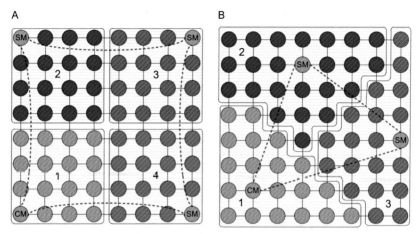

Fig. 20 8 × 8 HWNoC-based MCSoC (A) four regions and (B) three regions.

The WRs are interconnected to form a wireless highway if they fall within each other transmission range. Thus, in each dedicated region only one WR exists and the WR is associated to serve as the access point to the highway. For network efficiency, HWNoC is partitioned in a way that any core within a region has the minimum hop count toward the WR of that region than the WRs of the other regions. For borderline cases that a core may have the same hop count from two or more WRs, the core will be randomly assigned to one of the candidate regions. Fig. 20 shows two 64-core HWNoC-based MCSoCs with four and three regions.

Moreover, regardless of the number of regions, four Cartesian coordinate systems are defined as Down-Left (DL), Top-Left (TL), Top-Right (TR), and Down-Right (DR) shown in Fig. 21. Origin of each coordinate system is one of the four corners of the network. At each moment, there is one active region (*active_R*) along with one active coordinate system (*active_C*). Furthermore, the WRs are equipped with the control logic to manage the application mapping within their regions. One of the WRs is assigned as CM and the other WRs are named Regional Managers (RMs). Since each manager is responsible to assign the tasks on its own region, the hierarchical managing scheme helps balance the workload distribution between different managers.

3.2.2 Application Representation
In many embedded system applications like robotics, biomedical systems, and multimedia control systems, not only the tasks of each application are able to

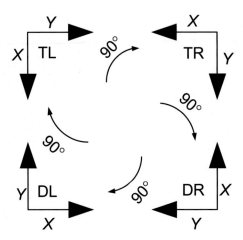

Fig. 21 Cartesian coordinate systems.

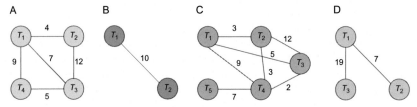

Fig. 22 Task graphs of different applications.

communicate with each other (i.e., intracommunication), but also multiple applications may also communicate with each other (i.e., intercommunication).

Intracommunication: An undirected graph, naming Task Graph (TG), represents each application and intracommunication between its tasks in the system. Each vertex denotes one task of the application, while each edge stands for communication between each two tasks as given in Eq. (7). The TGs of four applications each with two to five tasks are shown in Fig. 22. The amount of data transferred between any two tasks is indicated on the edge.

$$\forall t_i \in T, \forall e_{i,j} \in E, app = TG(T, E) \tag{7}$$

Intercommunication: Since the applications are considered to enter the system at runtime, no static graph can be defined for intercommunication between different applications. An incoming application may request to communicate with an already mapped application. Moreover, an existing application in the system may ask to communicate with a newly mapped one or the one which is not yet mapped onto the system. Thus, the intercommunication graph between different applications is highly dynamic.

3.2.3 Mapping Algorithm

RRM tries to map incoming applications region by region in a round–robin manner to balance the thermal distribution globally while periodically rotating the Cartesian coordinate system to balance the thermal distribution within each region locally. On the other hand, the tasks of each application are mapped with regard to the minimum Hop-Count Contiguity (HCC) in order to reduce congestion caused by long-distance communications of the same tasks of each application.

Fig. 23 represents the RRM algorithm. In *Initialization()* function, the number of regions (i.e., r), set of regions (i.e., $R = \{R_1, R_2, ..., R_r\}$), and set of coordinate systems (i.e., $C = \{DL, TL, TR, DR\}$) are initialized. Moreover, the active region and active coordinate system are also initialized to the first element of each set (i.e., $active\_R = R_1$ and $active\_C = DL$). Then, applications are chosen based on the FCFS policy since no background information is considered about incoming applications. In case of having background

> **r: Number of regions (i.e., number of WRs)**
> **C: Set of coordinate systems = {DL, TL, TR, DR}**
> **R: Set of regions = {R₁, R₂,..., Rᵣ}**
> **A: Set of applications**
> **active_C: Active coordinate system**
> **active_R: Active region**
>
> **Initialization() while *true* do**
> > **if *A is empty* then**
> > > **⌊ Sleep()**
> >
> > **app = Choose an application from A**
> > **free_Y = Choose the set of free cores with the smallest Y**
> > **free_XY = Choose the free core with the smallest X from free_Y**
> > **ft = FirstTask(app, free_XY)**
> > **Map(ft, free_XY)**
> > **ConMap(app, free_XY)**
> > **if *active_R == Rᵣ* then**
> > > **⌊ active_C = Choose the next coordinate system from C**
> >
> > **active_R = Choose the next region from R**

Fig. 23 RRM algorithm.

information about the incoming applications, an appropriate application selection policy can be applied which is beyond the scope of this chapter.

In each region, RRM first tries to find a set of free cores with the smallest "Y"s. Then among them, the core with the smallest "X" is chosen in order to map the first task of the application. The first task of each application is returned from the *FirstTask(app, free_XY)* function. This function returns the task of the selected application (i.e., *app*) with the equal or smaller number of edges than the available free cores around the chosen core (i.e., *free_XY*). If there is more than one task with the aforementioned criteria, then the first task would be the one with the most intensive communication among the candidates. If there is no task with the equal or smaller edges as the available free cores around *free_XY*, the task with the least intensive communication among all the tasks is chosen. In the case of existing two or more candidates with the same characteristics, one of them is randomly chosen.

Fig. 24 shows the first task selection for different values of available free cores around *free_XY* for the four applications of Fig. 22. For example, in application A (i.e., Fig. 22A) if the number of available free cores around *free_XY* is "1," there is no task with the equal or smaller number of edges as "1"; thus, the task with the least intensive communication among all the tasks of application A (i.e., T_4) is selected. However, if the available free cores are "2," two candidates (i.e., T_2 and T_4) have equal or smaller number of edges than "2"; among them, T_2 is selected because it has more intensive communications than T_4 (i.e., 16 vs 14). If the number of available cores is equal or greater than "3," the task with the most intensive communications (i.e., T_3) will be chosen. Note that in a mesh-based NoC the maximum available cores around each core are eight. Also in each region, only the free cores within that region are considered. The first task selection helps choose the most suitable central task of the selected application (i.e., *app*) to be mapped into the chosen free core of the system (i.e., *free_XY*) according to minimum HCC mapping.

After mapping the first task to *free_XY*, RRM tries to map the other tasks of the application based on minimum HCC around the first task within that

Number of free cores	0	1	2	3	4	5	6	7	8
Application A: Fig. 22A	T_4	T_4	T_2	T_3	T_3	T_3	T_3	T_3	T_3
Application B: Fig. 22B	T_1/T_2	T_1/T_2	T_1/T_2	T_1/T_2	T_1/T_2	T_1/T_2	T_1/T_2	T_1/T_2	T_1/T_2
Application C: Fig. 22C	T_5	T_5	T_5	T_3	T_4	T_4	T_4	T_4	T_4
Application D: Fig. 22D	T_2	T_3	T_1	T_1	T_1	T_1	T_1	T_1	T_1

Fig. 24 First task selection example.

region. Minimum HCC is defined as minimum overall hop count between all the cores in which the application is mapped into. In the case that the application does not fit into the current region (i.e., *active_R*), the current region will be merged with the next region temporarily. After mapping of each application, the active region is shifted to the next region to balance the thermal distribution globally. Moreover, after a complete round (i.e., all the regions become active once in one coordinate system) the origin of the coordinate system is rotated to the next origin to balance the thermal distribution within each region as well. Note that when the RRM algorithm reaches the last element of R (or C), it starts from the beginning again, i.e., *active_R* is set to R_1 (or *active_C* is set to DL). In the case that there is no available application to be mapped into the system, RRM goes to the *Sleep()* mode until a new application arrives and signals to wake up.

Overall, RRM tries to map the task of each application as contiguous as possible based on minimum HCC to avoid long-distance communications that mostly influence the WRs. Also, it tries to spread the temperature across the chip by periodically changing the regions (i.e., global heat distribution) and coordinate systems (i.e., local heat distribution).

3.2.4 Step-by-Step Examples

In order to have a better understanding of RRM algorithm, two step-by-step examples are discussed as follows. In each example the four applications of Fig. 22 as well as four more applications are considered to be mapped into the system. Fig. 25 represents a visual example of the RRM algorithm in a region-based 64-core HWNoC with four symmetric regions. In Fig. 25A application A (i.e., Fig. 22A) with four tasks arrives. Since the first region is active and the active coordinate is DL (i.e., down-left), RRM chooses the core $(0, 0)$ as *free_XY* because it is the core with the smallest Y and then the smallest X in the first region based on DL coordinate system. Since the number of available free cores around the core $(0, 0)$ is "3," T_3 is chosen as the first task of application A based on Fig. 24. Then T_3 is assigned to core $(0, 0)$. Afterward, the other three tasks of application A are mapped as contiguous as possible to T_3. Now the active region is shifted to the second one.

In Fig. 25B application B (i.e., Fig. 22B) with two tasks arrives. The second region is active and the active coordinate is still DL. Thus, RRM chooses the core $(0, 4)$ as *free_XY* because it is the core with the smallest Y and then the smallest X in the second region with respect to DL coordinate system. Then, either T_1 or T_2 is selected as the first task of application B according to Fig. 24. Since application B has only one more task, it can be mapped at core $(1, 4)$ or core $(0, 5)$ with respect to contiguity. In these cases

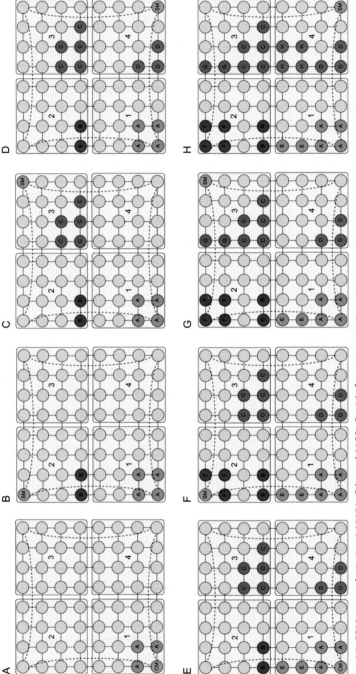

Fig. 25 RRM example in 8 × 8 HWNoC-based MCSoC with four symmetric regions.

one of the cores is randomly chosen as the target. Then the active region is changed to the third one.

In Figs. 24D and 25C applications C (i.e., Fig. 22C) and D (i.e., Fig. 22D) are mapped. Then, because the active region is reached to the last region in the first round, the coordinate system is rotated to TL (i.e., top-left) and the active region is again shifted to the first one. Now, in Fig. 25E for application E with four tasks the core $(4, 0)$ of new coordinate system (i.e., TL) is chosen as *Free_XY*. The procedure continues for the applications F, G, and H.

Moreover, Fig. 26 shows a visual example of RRM algorithm in a region-based 64-core HWNoC-based MCSoC with three asymmetric regions. The RRM works exactly the same for both symmetric and asymmetric regions. However, more attention is required to following up the step-by-step procedure in asymmetric regions. At first, in Fig. 26A application A (i.e., Fig. 22A) with four tasks arrives. Since the first region is active and the active coordinate is DL, RRM chooses the core $(0, 0)$ as *free_XY*. Since the number of available free cores around the core $(0, 0)$ is "3," T_3 is chosen as the first task of application A based on Fig. 24. Then T_3 is assigned to core $(0, 0)$. After mapping the other tasks of application A as contiguous as possible to T_3, the active region is shifted to the second one.

In Fig. 26B application B (i.e., Fig. 22B) with two tasks arrives. Since the second region is active and the active coordinate is still DL, RRM chooses the core $(3, 3)$ because it is the core with the smallest Y and then the smallest X in the second region based on the DL coordinate system. Then, either T_1 or T_2 is selected as the first task of application B according to Fig. 24. Since application B has only one more task, it can be only mapped at the core $(3, 4)$ with respect to the minimum HCC. Then the active region is shifted to the third one.

In Fig. 26C application C (i.e., Fig. 22C) with five tasks arrives and core $(6, 0)$ is selected as *free_XY*. Since the number of available free cores around the core $(6, 0)$ is "4," T_4 is chosen as the first task of application C based on Fig. 24.

After mapping application C, since all the regions are met in the first round, the coordinate system is rotated to TL. Applications D, E, and F are mapped based on the new coordinate system to the first, second, and third regions, respectively. Next, because all the regions are met again, the coordinate system is rotated to TR (i.e., top-right). Now, as shown in Fig. 26G for application G with two tasks, the core $(5, 4)$ is chosen as *free_XY* because it has the smallest Y and then the smallest X among all the free cores in the first region with respect to the TR coordinate system. The active region is then shifted to the second one and the procedure is repeated for application H.

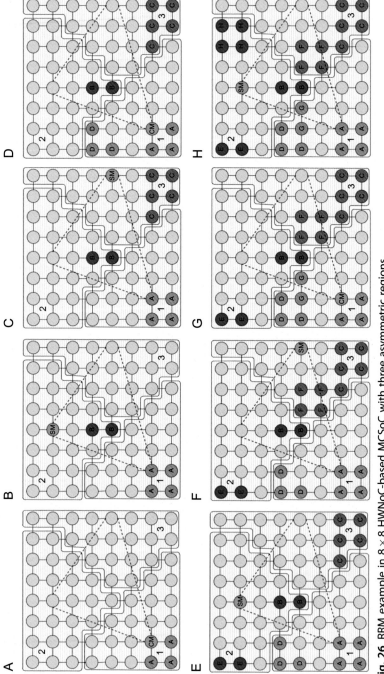

Fig. 26 RRM example in 8×8 HWNoC-based MCSoC with three asymmetric regions.

3.3 Methodology

The basic experimental setups are the same as Section 2.3. In addition, several sets of applications each with 2–5 tasks are generated using TGG [32] where the amount of data transferred from the source task to the destination task are randomly distributed between 2 and 36 flits of data. Each application is considered 75% intracommunication and 25% intercommunication between the other applications. The intercommunication between different applications is conducted by the first task of each application mapped to the system.

Applications are scheduled based on the FCFS policy and the maximum possible scheduling rate is λ_{full}. An allocation request for the scheduled application is sent to the CM of the system. CM then based on the active region sends the information to the responsible RM through the hierarchical managing network. The hierarchical XY routing algorithm taken from Ref. [23] is implemented in which intraregion communications are handled through wired path and interregion communications are supported via both wired and wireless paths. Two 64-core HWNoC (52% dark silicon) with three and four WRs (Fig. 20) are considered in the simulations. Comparisons are also made between RRM and random task mapping as baseline in addition to the congestion-aware dynamic task mapping algorithm (DMA) presented in Ref. [31].

3.4 Experimental Results

3.4.1 Hop Counts and Energy Saving

As shown in Refs. [33,34], decreasing Manhattan distance (MD) between tasks of application edges is an effective way to minimize the communication energy consumption of the applications. The percentage of packets that are delivered over different path lengths (i.e., MD) is illustrated in Fig. 27. The experiments have been run for different algorithms in the injection rate of 0.5 λ_{full}. As can be seen, more than 60% (and more than 50%) of the packets are delivered by one-hop distance using RRM algorithm in 64-core HWNoC with four regions (and three regions).

Accordingly, Fig. 28 represents the average MD for different algorithms in the injection rate of 0.5 λ_{full} based on the percentage of the intracommunication and intercommunication. By decreasing the percentage of intercommunication between applications, more energy can be saved by RRM, since it maps all the tasks of each application based on minimum HCC. As can be seen RRM outperforms DMA in less than 5% intercommunication between different applications.

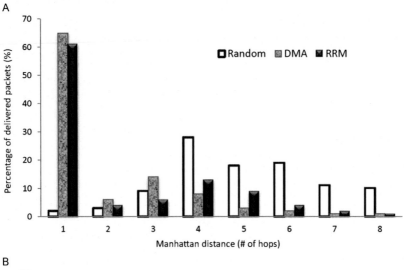

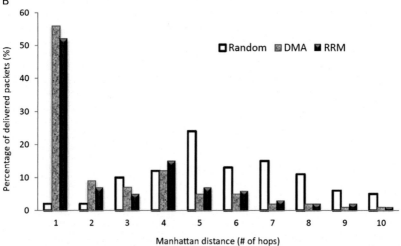

Fig. 27 Percentage of delivered packets in different path lengths in 64-core HWNoC (A) four regions and (B) three regions.

Intracommunication (%)/intercommunication (%)		70/30	75/25	80/20	85/15	90/10	95/5	100/0
64-core HWNoC with four regions	Random	5.01	5.08	5.21	5.36	5.14	5.2	5.07
	DMA	2.34	1.93	1.61	1.31	1.13	0.93	0.78
	RRM	2.78	2.3	1.84	1.55	1.27	0.92	0.66
64-core HWNoC with three regions	Random	5.61	5.79	5.6	5.87	6.03	5.92	5.72
	DMA	2.98	2.47	2.07	1.68	1.4	1.16	0.96
	RRM	3.21	2.77	2.2	1.82	1.49	1.09	0.78

Fig. 28 Average MD comparison.

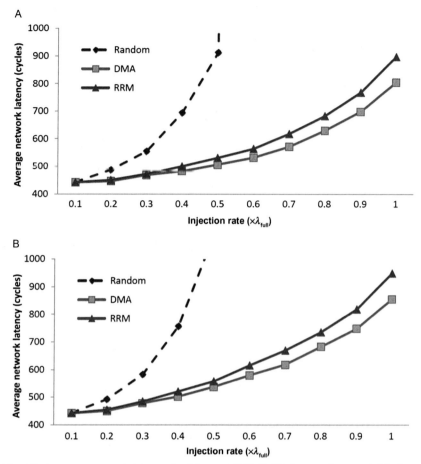

Fig. 29 Average network latency in 64-core HWNoC (A) four regions and (B) three regions.

3.4.2 Network Latency and Congestion Avoidance

Fig. 29 shows the average network latency for different algorithms. It is supposed that there is no gap between application arrivals. As can be seen, RRM has a reasonable average network latency next to DMA. Contiguous mapping of each application tasks in both DMA and RRM results in lower network latency than random mapping. Also, in DMA and RRM, by increasing the injection rate, the network becomes uniformly congested because the usage of WRs is more balanced.

It can be concluded from Figs. 27 and 29 that employing hierarchical XY routing that uses wired paths for short-distance communications combined

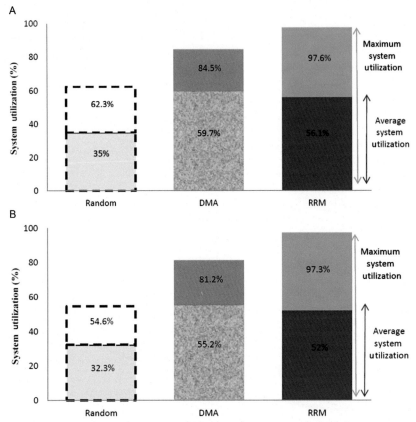

Fig. 30 System utilization in 64-core HWNoC (A) four regions and (B) three regions.

with RRM can dynamically reduce the congestion over WRs while minimized the communication energy consumption.

3.4.3 System Utilization

System utilization is another parameter that has been analyzed among the different algorithms. As shown in Fig. 30, RRM has lower average system utilization than DMA mapping but has better maximum system utilization that is defined as the highest percentage of the utilization during the simulation time. Note that the system utilization is based on the number of tasks can be mapped on nondark cores (i.e., 48% of the cores in 64-core HWNoC) which communicate with each other without dropping due to the high congestion.

This happens because DMA tries to map not only all the tasks of each application but also all the applications contiguous (i.e., close to CM) that results in better average system utilization. On the other hand, RRM unlike DMA does not suffer from area fragmentation and can reach higher maximum utilization (i.e., almost 98%).

3.4.4 Thermal Analysis

Figs. 31 and 32 demonstrate the thermal distribution of different algorithms in the maximum system utilization based on Fig. 30. Furthermore, Fig. 33 shows the average and the peak temperature comparisons. Permanent hot spots in the system greatly increase the failure probability. Since in dark silicon age more than half of the chip is dark, RRM utilizes the dark cores in order to efficiently avoid hot spots in the system. On the other hand, DMA suffers from severe hot spot around CM and random mapping has multiple hot spots around WRs. Moreover, unlike RRM, the peak temperature is gotten worse in DMA by decreasing the number of WRs.

3.5 Summary

In this section, a temperature- and congestion-aware task mapping algorithm named RRM was introduced in order to solve some of the key concerns in future HWNoC-based MCSoCs. Simulation results showed significant improvement in both congestion and temperature control of the system. More than 60% (and more than 50%) of the packets are delivered by one-hop distance using RRM algorithm in 64-core HWNoC with four regions (and three regions); this saves communication energy consumption significantly. RRM also has a reasonable average network latency. Contiguous mapping of each application tasks results in lower network latency and finally total execution time gain. Overall, RRM provides more than 50% system utilization and about 98% maximum system utilization. Moreover, the heat is distributed evenly across the whole chip using RRM algorithm. The peak and average temperature are less than 352 and 330 K, respectively, in 64-core HWNoC. Fig. 34 shows the comparison summary of different mapping algorithms.

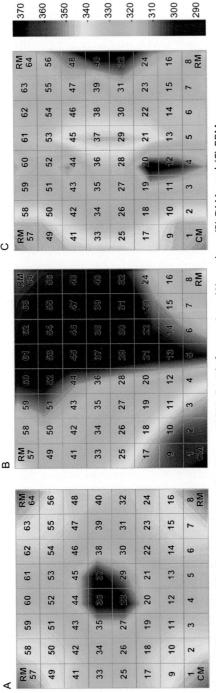

Fig. 31 Thermal distribution comparison in 64-core HWNoC with four regions (A) random, (B) DMA, and (C) RRM.

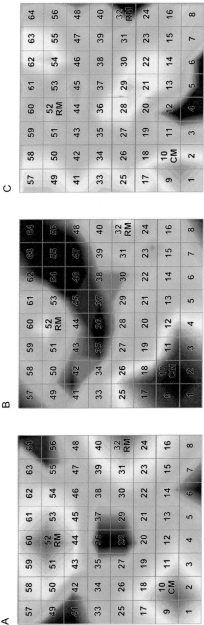

Fig. 32 Thermal distribution comparison in 64-core HWNoC with three regions (A) random, (B) DMA, and (C) RRM.

		Random	DMA	RRM
64-core HWNoC with four regions	Average temperature (K)	333.6	318.1	329.6
	Peak temperature (K)	356.4	372.8	351.7
64-core HWNoC with three regions	Average temperature (K)	330.1	323.5	329.9
	Peak temperature (K)	357.1	375.2	351.4

Fig. 33 Average and peak temperature comparison.

	Random mapping	Dynamic mapping algorithm	Round rotary mapping
Energy consumption	High communication energy consumption	Low communication energy consumption	Low communication energy consumption
Execution time	High network latency	Low network latency	Low network latency
System utilization	Medium system utilization	High average system utilization	High average and maximum system utilization
Temperature	Hot spots around WRs	Intensive hot spot around CM	No hot spot

Fig. 34 Comparison summary.

4. CONCLUSION AND THE FUTURE OUTLOOK

Due to the utilization wall problem, the threshold voltage cannot be scaled without exponentially increasing leakage, and as a result, the operating voltage should be kept roughly constant. This is an exponentially worsening problem that accumulates with each process generation [35]. Recent studies [1] have predicted that, on average, 52% of a chip's area will stay dark for the 8-nm technology node.

Based on the multiobjective nature of MCSoCs in dark silicon age, the many-core system should be capable of observing and automatically adapting its behavior and recourses in order to accomplish its assigned job efficiently. On-chip self-awareness is a key enabling technology for efficient use of heterogeneous architectures and applications with guaranteed runtime system objectives [36].

Moreover, we are expecting billions of devices connected to the Internet of Things in the near future. Manual management to reach multiobjective goals of these devices will soon become impossible. Thus, we need to bring learning-based methods and control theory into the game as well. As a result, the technology trend in embedded system design is not only shifting from single-objective homogenous designs to multiobjective heterogeneous ones but also inevitably changing from manually tune-up approaches to self-aware platforms.

REFERENCES

[1] J. Henkel, H. Khdr, S. Pagani, M. Shafique, New trends in dark silicon, in: ACM/ EDAC/IEEE Design Automation Conference (DAC), 2015. Article 119.

[2] N. Goulding-Hotta, J. Sampson, G. Venkatesh, S. Garcia, J. Auricchio, P.-C. Huang, M. Arora, S. Nath, V. Bhatt, J. Babb, S. Swanson, M. Taylor, The GreenDroid mobile application processor: an architecture for silicon's dark future, IEEE Micro 31 (2) (2011) 86–95.

[3] N. Hardavellas, M. Ferdman, B. Falsafi, A. Ailamaki, Toward dark silicon in servers, IEEE Micro 31 (4) (2011) 6–15.

[4] A. Raghavan, Y. Luo, A. Chandawalla, M. Papaefthymiou, K.P. Pipe, T.F. Wenisch, M.M.K. Martin, Computational sprinting, in: IEEE International Symposium on High-Performance Computer Architecture (HPCA), 2012, pp. 1–12.

[5] L. Benini, G. De Micheli, Networks on chips: a new SoC paradigm, IEEE Comput. 35 (1) (2002) 70–78.

[6] J. Zhan, Y. Xie, G. Sun, NoC-sprinting: interconnect for fine-grained sprinting in the dark silicon era, in: ACM/EDAC/IEEE Design Automation Conference (DAC), 2014. Article 160.

[7] A. Rezaei, D. Zhao, M. Daneshtalab, H. Wu, Shift sprinting: fine-grained temperature-aware NoC-based MCSoC architecture in dark silicon age, in: ACM/EDAC/IEEE Design Automation Conference (DAC), 2016. Article 155.

[8] Y. Zhang, L. Peng, F. Xin, H. Yue, Lighting the dark silicon by exploiting heterogeneity on future processors, in: ACM/EDAC/IEEE Design Automation Conference (DAC), 2013. Article 82.

[9] B. Raghunathan, Y. Turakhia, S. Garg, D. Marculescu, Cherry-picking: exploiting process variations in dark-silicon homogeneous chip multi-processors, in: Design, Automation & Test in Europe Conference & Exhibition (DATE), 2013, pp. 39–44.

[10] ITRS, International Technology Roadmap for Semiconductors, 2013. edition.

[11] M. Shafique, S. Garg, J. Henkel, D. Marculescu, The EDA challenges in the dark silicon era: temperature, reliability, and variability perspectives, in: ACM/EDAC/IEEE Design Automation Conference (DAC), 2014. Article 185.

[12] S. Pagani, H. Khdr, W. Munawar, J.-J. Chen, M. Shafique, M. Li, J. Henkel, TSP: thermal safe power: efficient power budgeting for many-core systems in dark silicon, in: International Conference on Hardware/Software Codesign and System Synthesis (CODES), 2014. Article 10.

[13] J.M. Allred, S. Roy, K. Chakraborty, Long term sustainability of differentially reliable systems in the dark silicon era, in: International Conference on Computer Design (ICCD), 2013, pp. 70–77.

[14] B. Raghunathan, S. Garg, Job arrival rate aware scheduling for asymmetric multi-core servers in the dark silicon era, in: International Conference on Hardware/Software Codesign and System Synthesis (CODES), 2014. Article 14.

[15] H. Khdr, S. Pagani, M. Shafique, J. Henkel, Thermal constrained resource management for mixed ILP-TLP workloads in dark silicon chips, in: ACM/EDAC/IEEE Design Automation Conference (DAC), 2015. Article 179.

[16] M. Shafique, D. Gnad, S. Garg, J. Henkel, Variability-aware dark silicon management in on-chip many-core systems, in: Design, Automation & Test in Europe Conference & Exhibition (DATE), 2015, pp. 387–392.

[17] N. Kapadia, S. Pasricha, VARSHA: variation and reliability-aware application scheduling with adaptive parallelism in the dark-silicon era, in: Design, Automation & Test in Europe Conference & Exhibition (DATE), 2015, pp. 1060–1065.

[18] A. Rezaei, F. Safaei, M. Daneshtalab, H. Tenhunen, HiWA: a hierarchical wireless network-on-chip architecture, in: IEEE International Conference on High Performance Computing & Simulation (HPCS), 2014, pp. 499–505.

[19] A. Rezaei, D. Zhao, M. Daneshtalab, H. Zhou, Multi-objective task mapping approach for wireless NoC in dark silicon age, in: IEEE Euromicro International Conference on Parallel, Distributed and Network-Based Computing (PDP), 2017, pp. 589–592.

[20] H. Esmaeilzadeh, A. Sampson, M.R. ingenburg, L. Ceze, D. Grossman, D. Burger, Addressing dark silicon challenges with disciplined approximate computing, in: Dark Silicon Workshop (DaSi) With International Symposium on Computer Architecture (ISCA), 2012, pp. 1–2.

[21] J. Cong, B. Xiao, Optimization of interconnects between accelerators and shared memories in dark silicon, in: International Conference on Computer-Aided Design (ICCAD), 2013, pp. 630–637.

[22] K. Swaminathan, E. Kultursay, V. Saripalli, V. Narayanan, M.T. Kandemir, S. Datta, Steep-slope devices: from dark to dim silicon, IEEE Micro 33 (5) (2013) 50–59.

[23] A. Rezaei, M. Daneshtalab, F. Safaei, D. Zhao, Hierarchical approach for hybrid wireless network-on-chip in many-core era, Elsevier Int. J. Comput. Electr. Eng. 51 (C) (2016) 225–234.

[24] B. Goodarzi, H. Sarbazi-Azad, Task migration in mesh NoCs over virtual point-to-point connections, in: IEEE Euromicro International Conference on Parallel, Distributed and Network-Based Processing (PDP), 2011, pp. 463–469.

[25] V. Catania, A. Mineo, S. Monteleone, M. Palesi, D. Patti, Noxim: an open, extensible and cycle-accurate network on chip simulator, in: IEEE International Conference on Application-Specific Systems, Architectures and Processors (ASAP), 2015, pp. 162–163.

[26] L. Wang, K. Skadron, Dark vs. Dim Silicon and Near-Threshold Computing Extended Results, Technical Report (UVA-CS-2013-01), Department of Computer Science, University of Virginia, 2013.

[27] W. Huang, S. Ghosh, S. Velusamy, K. Sankaranarayanan, K. Skadron, M. Stan, HotSpot: a compact thermal modeling methodology for early-stage VLSI design, IEEE Trans. Very Large Scale Integr. VLSI Syst. 14 (5) (2006) 501–513.

[28] C. Bienia, S. Kumar, J.P. Singh, K. Li, The PARSEC benchmark suite: characterization and architectural implications, in: International Conference on Parallel Architectures and Compilation Techniques (PACT), 2008, pp. 72–81.

[29] J.A. Butts, G.S. Sohi, A static power model for architects, in: IEEE/ACM International Symposium on Microarchitecture (MICRO), 2000, pp. 191–201.

[30] A. Rezaei, M. Daneshtalab, D. Zhao, M. Modarressi, SAMi: self-aware migration approach for congestion reduction in NoC-based MCSoC, in: IEEE International System-on-Chip Conference (SOCC), 2016.

[31] A. Rezaei, M. Daneshtalab, D. Zhao, F. Safaei, X. Wang, M. Ebrahimi, Dynamic application mapping algorithm for wireless network-on-chip, in: IEEE Euromicro International Conference on Parallel, Distributed and Network-Based Computing (PDP), 2015, pp. 421–424.

[32] Source Forge, Task Graph Generator (TGG), Available: http://sourceforge.net/projects/taskgraphgen/.

[33] C.L. Chou, U.Y. Ogras, R. Marculescu, Energy- and performance-aware incremental mapping for networks on chip with multiple voltage levels, IEEE Trans. Comput. Aided Des. Integr. Circuits Syst. 27 (10) (2008) 1866–1879.

[34] A. Rezaei, M. Daneshtalab, M. Palesi, D. Zhao, Efficient congestion-aware scheme for wireless on-chip networks, in: IEEE Euromicro International Conference on Parallel, Distributed and Network-Based Computing (PDP), 2016, pp. 742–749.
[35] M.B. Taylor, Is dark silicon useful? Harnessing the four horsemen of the coming dark silicon apocalypse, in: ACM/EDAC/IEEE Design Automation Conference (DAC), 2012, pp. 1131–1136.
[36] N. Dutt, A. Jantsch, S. Sarma, Toward smart embedded systems: a self-aware system-on-chip (SoC) perspective, ACM Trans. Embed. Comput. Syst. 15 (2) (2016). Article 22.

ABOUT THE AUTHORS

Amin Rezaei is currently a Ph.D. candidate in Electrical Engineering and Computer Science at Northwestern University, the USA. He received his B.Sc. degree in Computer Engineering from University of Isfahan, Iran, in 2011 and two M.Sc. degrees one in Computer Engineering from Shahid Beheshti University, Iran, in 2014 and the other in Computer Science from University of Louisiana at Lafayette, the USA, in 2016. His main research interests include Parallel Computing, Dark Silicon, and Logic Encryption.

Masoud Daneshtalab is currently is currently a tenured associate professor at Mälardalen University (MDH) and visiting researcher at Royal Institute of Technology (KTH), Sweden. Before that he was lecturer and project manager at University of Turku in Finland. He has been appointed as Associate Editors of Elsevier Journals of Computers & Electrical Engineering and Microprocessors and Microsystems; and in the Editorial Board of The Scientific World Journal, IJDST, IJARAS, IJERTCS, and IJDATICS. He is also in the Euromicro's board of directors since 2016 while representing Sweden in the management committee of the EU COST Actions IC1202 (TACLe). His research interests include interconnection networks, many-core and reconfigurable systems, and neuromorphic architecture. He has published 1 book, 4 book chapters, and over 200 refereed international journals and conference papers. He is currently TPC member of different IEEE and ACM conferences, including NOCS, DATE, ASPDAC, ICCAD, ESTIMedia, VLSI Design, ICA3PP, SOCC, VDAT, DSD, PDP, ICESS, Norchip, MCSoC, CADS, EUC, DTIS, NESEA, CASEMANS, NoCArc, MES, HPIN, PACBB, MobileHealth, and JEC-ECC.

Hai Zhou is an associate professor in Electrical Engineering and Computer Science at Northwestern University. He got his Ph.D. degree in Computer Sciences from the University of Texas at Austin, and his B.S. and M.S. degrees in Computer Science and Technology from Tsinghua University in Beijing, China. He was a recipient of a CAREER Award from the National Science Foundation. His research interests include VLSI computer-aided design, algorithm design, and formal methods. He has published more than 150 papers in flagship journals and conferences in these areas.

CHAPTER FOUR

Dark Silicon Aware Resource Management for Many-Core Systems

Heba Khdr*, Santiago Pagani†, Muhammad Shafique‡, Jörg Henkel*

*Chair for Embedded Systems (CES), Karlsruhe Institute of Technology (KIT), Karlsruhe, Germany
†ARM, Cambridge, United Kingdom
‡Institute of Computer Engineering, Vienna University of Technology (TU Wien), Vienna, Austria

Contents

Abstract

As Dennard's scaling stops mainly due to supply voltage limits, power densities rapidly increase on the chip. Hence, a significant amount of on-chip resources needs to stay

Advances in Computers, Volume 110
ISSN 0065-2458
https://doi.org/10.1016/bs.adcom.2018.03.002

dark, i.e., power-gated, in order to avoid thermal emergencies. This phenomenon is known in the literature as dark silicon. Conventional resource management techniques allocate the cores of the chip to the applications, without considering the dark silicon phenomenon. In this chapter, we discuss a dark silicon aware resource management technique that aims at maximizing the overall system performance under a temperature constraint. To achieve its goal, this technique determines the number of active (power-on) cores that should be allocated to each application and the voltage and frequency level of these cores. Additionally, it takes into account the instruction-level parallelism (ILP) and the thread-level parallelism (TLP) of the applications during its decision-making process. Moreover, the presented technique selects the positioning of dark cores such that they facilitate dissipating the generated heat on the active cores. That, in turn, reduces the temperature of the active cores and might allow to increase their voltage and frequency levels leading to further performance improvement. The evaluation of this technique shows its ability to improve the performance with an average of 34% compared to a state-of-the-art technique of thermal-aware performance maximization.

1. INTRODUCTION

Due to the technology scaling and high integration of more transistors into smaller areas, power densities on the chip are continuously increasing, as shown in Fig. 1. Excessive power densities on the chip without a sufficient (costly) cooling system result in thermal emergencies. To mitigate this, some cores need to be power-gated (i.e., *dark* cores), while other cores can still be power-on (i.e., *active* cores) and run at different speeds (i.e., different voltage and frequency (v/f) levels).

Analyses of the amount of the dark cores on the chip for different technology nodes have been presented in [4, 5] based on the technology data from ITRS [2] and Intel [3]. According to [4], around 80% of a chip's area will stay dark for the 8 nm technology node. While the analysis presented in

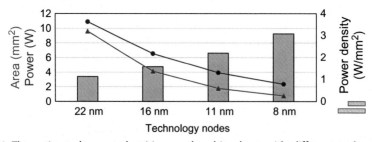

Fig. 1 The estimated power densities on the chip along with different technology nodes. The power consumptions and the area are estimated using McPAT [1] and based on the technology data from ITRS [2] and Intel [3].

[5] for the same technology node, i.e., 8 nm, shows that, on average, 52% of a chip's area will stay dark. Hence, this so-called *dark silicon* phenomenon is expected to be dominant in the future technology nodes.

Motivated by such analyses, several resource management techniques (e.g., [6–9]) have been proposed to maximize the system performance taking into account the dark silicon phenomenon. Most of these techniques suggest modeling dark silicon as a function of power constraint. That implies the number of active cores and their v/f levels are chosen such that the defined power constraint is not exceeded. The most commonly used power constraint in the literature is the thermal design power (TDP), given that the cooling system on the chip is designed to dissipate TDP. However, restricting the system to run below TDP is not enough to avoid thermal violations, as demonstrated in [10, 11]. An example about this will also be shown in Section 1.1. When a thermal violation occurs, that is, when the temperatures of some cores of the chip exceed a specified thermal threshold, defined as T_{critical}, dynamic thermal management (DTM) on the chip will be triggered to cool down the cores. For this purpose, DTM throttles down the v/f levels and it might even power-gate some cores. Triggering DTM will degrade the system performance, but it is necessary to save the chip from undesired overheating. In order to mitigate triggering DTM and its unpredictable performance losses, resource management techniques need to consider the temperature constraint (T_{critical}) rather than considering only the power constraint (TDP). Few dark silicon aware techniques (e.g., [12, 13]) consider the temperature constraint while mapping the applications to the chip. The main limitation of these techniques is that, they do not control the v/f levels of the cores, although the v/f levels play a significant role in determining the amount of dark silicon as well as the performance of the system.

Therefore, there is a need to propose a dark silicon aware resource management technique that determines both the number of active cores and their v/f levels considering the temperature constraint, while aiming at maximizing the system performance.

However, proposing such a technique can arise the following challenges: first, the same number of active and dark cores can lead to different on-chip temperatures according to their locations of the dark cores and the v/f levels of the active cores. Second, the resulting improvement of performance when

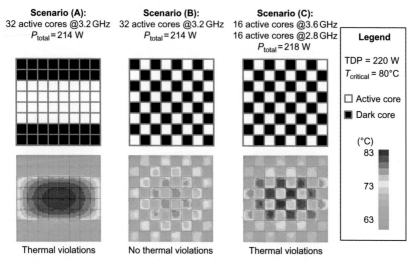

Fig. 2 Thermal violations might occur under TDP constraint depending on the locations of the dark cores and the v/f levels of the active cores.

increasing the number of active cores or increasing their v/f levels differs from an application to another according to the thread–level parallelism (TLP) and the instruction–level parallelism (ILP) of the application. Examples about these challenges are discussed in the following case studies.

1.1 Case Study 1: Impact of the Locations of Dark Cores and the v/f Levels of Active Cores on Temperature

In this case study, we present three scenarios of thermal simulations on a homogeneous 64-core chip.[a] The critical temperature $T_{critical}$ of this chip is equal to 80°C. The employed chip has a TDP of 220 W, where all three scenarios consider TDP as the power constraint of the chip, i.e., the total power consumption must be kept below TDP. Thus, in all three scenarios, we execute four instances of an x264 application from the PARSEC benchmark suite [15], each instance running 8 parallel dependent threads, which gives in total 32 active cores.

In the first scenario, all active cores run at 3.2 GHz. The application threads are mapped close to each other, as shown in Fig. 2A. We can notice from the thermal map in Fig. 2A that, the resulting temperatures of several

---

[a] Alpha 21264 cores in 22 nm, simulated with gem5 [14] and McPAT [1]. Further details are presented in Section 6.1.

cores violate the critical temperature $T_{critical}$. In the second scenario, all active cores run at the same selected frequency in the first scenario, i.e., 3.2 GHz. However, the application threads are mapped to the active cores according to a chessboard pattern, as shown in Fig. 2B. We can observe from the resulting thermal map that no thermal violation occurs, although the number of active cores and their v/f level are the same as the first scenario. The reason is that the dark cores in this scenario are distributed around the active cores, and thereby the generated heat on the active cores is better dissipated, and thereby, their temperatures are decreased. Thus, the first and the second scenarios show that the distribution of the active and dark cores on the chip affects the thermal profile, even though the active cores are consuming the same amount of power.

In the third scenario (Fig. 2C), the locations of active and dark cores are determined similar to the second scenario, but the v/f levels of the active cores are differently set. Namely, in this scenario 16 cores run at 3.6 GHz and the other 16 cores run at 2.8 GHz. The cores of the lower frequency are near the chip edges, while the ones of the higher frequency are close to each other and near the center of the chip. Although the total power consumption in this scenario is also below TDP and the dark cores are distributed around the active cores, we notice that thermal violations occur on the centralized cores, due to their high v/f level that results in high power densities.

In summary, three important observations can be made from this case study:

(1) Although the total power consumption in all three scenarios is very similar and below TDP, thermal violations occur in two scenarios. This again shows that using only a power constraint by resource management is not enough to avoid thermal violations.

(2) The locations of dark cores can be determined through mapping decisions such that they facilitate dissipating the generated heat on the active cores.

(3) The v/f levels of the active cores should be determined together with the mapping decisions in order to avoid thermal violations.

Thus, to avoid thermal violations and their associated performance losses, dark silicon aware resource management needs to determine the number of active/dark cores and the v/f levels of the active cores based on temperature constraint rather than TDP constraint. Moreover, the positioning of dark cores can be leveraged to dissipate more heat from active cores.

1.2 Case Study 2: Impact of TLP and ILP on Performance

As discussed earlier, resource management needs to determine the number of active cores and their v/f levels in order to maximize the system performance under the temperature constraint. In many-core systems, there can be multiple applications that need to be mapped and run on the chip at the same time. Therefore, resource management should determine the number of active cores that should be assigned to each application and the v/f level of its cores,[b] such that the overall system performance is maximized. However, according to Amdahl's law the resulting performance gain of assigning multiple cores to an application depends on the part of the application that can be parallelized. In other words, it depends on the TLP of the application, and it differs from one application to another. The ILP denotes how many instructions of the application can be executed simultaneously. When the application has higher ILP, its performance can be significantly increase by increasing the v/f level of the processor. Hence, the resulting performance gain of increasing the v/f level of the cores of an application differs from one application to another, based on the ILP of the application.

To analyze the impact of the TLP and ILP of the applications on their performance, we show in Fig. 3 the resulting execution time of two applications, i.e., blackscholes and ferret [15]; first when increasing the number of cores at a fixed v/f level, and secondly when increasing the v/f levels at a fixed number of cores. Less execution time means more performance gain. As shown in Fig. 3, the first application, blackscholes, gains more performance when the number of cores is increased compared to its performance when the v/f level of its cores is increased. Contrarily, ferret application gains more performance when the v/f level of its cores is increased.

Therefore, to increase the performance of a high-ILP application, dark silicon aware resource management can increase the amount of dark cores in close vicinity to the active cores of that application, allowing to increase the v/f level of its active cores (as some of the incurred heat will be dissipated by the neighboring dark cores). Contrarily, the performance of a high-TLP application increases by assigning more active cores to it with lower v/f level, compared to high-ILP applications.

---

[b] The cores that are assigned to a multithreaded application should run at the same v/f level in order to avoid synchronization stalls between its threads.

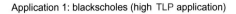

Application 1: blackscholes (high TLP application)

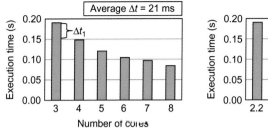

→ Average performance gain of increasing the number of cores is 82% more
than the resulting gain of increasing the v/f level

Application 2: ferret (high ILP application)

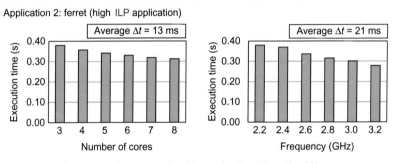

→ Average performance gain of increasing the v/f level is 58% more
than the resulting gain of increasing the number of cores

Fig. 3 The resulting execution time of two applications that have different TLP and ILP, along with increasing either the number of cores assigned to them or the v/f level of their cores.

1.3 Summary of the Technique Presented in this Chapter

In this chapter, we discuss a dark silicon aware resource management technique, called *DsRem* [16], that solves the aforementioned challenges. The goal of *DsRem* is to improve the overall system performance in many-core chips under the temperature constraint, $T_{critical}$. To achieve its goal, *DsRem* needs to select the number of active and dark cores, their locations, and the v/f levels of the active cores through determining the number of threads of each application and the thread-to-core assignment as the following steps show

- Initially, *DsRem* finds the *optimal* number of cores (threads) that should be allocated to each application jointly with their *optimal* v/f levels (considering the TLP and ILP of the applications) such that the overall system performance is maximized under the TDP constraint.

- Based on the initial resource allocation, *DsRem* determines the application mapping on the chip and selects the locations of the dark cores such that the temperatures of the active cores are reduced.
- After determining the application mapping, *DsRem* estimates the steady-state temperatures of the cores that would be incurred if the resource allocation and application mapping determined by the first and second steps were applied. According to the estimated temperatures, *DsRem* modifies the resource allocation, i.e., the number of cores and the v/f levels, in order to avoid thermal violations or to exploit available thermal headroom.[c] That implies, *DsRem* removes the unsafe TDP constraint assumption at this step; rather it considers the temperature constraint.

2. STATE-OF-THE-ART RESOURCE MANAGEMENT TECHNIQUES

Many resource management techniques have emerged to manage the resources of the chip using different policies and considering different constraints. An early approach presented in [17] proposed hardware/software partitioning for the processor such that a low power system can be obtained. However, due to the emergence of multi- and many-core systems, the focus of research has shifted to propose techniques that partition/distribute the cores among multiple applications.

Some of these techniques (e.g., [18–22]) focus on distributing the cores of the chip among the competing applications aiming at maximizing the overall system performance. For example, the work proposed in [18] distributes the cores among the applications based on the speedup curve of each application. More cores will be assigned to the applications who have higher speedup gain. Similar work is presented in [19], but additionally the heterogeneity of the cores is considered. In [20], an adaptive clustering policy is proposed to manage the resources among applications. Each application is assigned to one virtual cluster, such that the cluster adapts its size at runtime to increase the performance of its application. The proposed policy in [21] maps the applications to the chip, such that the fragmentation of the chip is reduced as much as possible. Reducing the fragmentation allows better mapping for new arriving applications, and thereby increases the system performance. However, none of these techniques considers power or temperature constraints.

---

[c] Thermal headroom denotes the difference between $T_{critical}$ and the core temperature.

Some techniques have been proposed to maximize the performance under a power constraint. For example, the work of [23] has proposed an incremental mapping algorithm that maps the applications to the cores such that the used power in communication between the application tasks over NoC is reduced. However, the computation power on the cores composes the majority of the total power consumption on the chip. Therefore, the technique proposed in [24] manages the chip resources such that the total power consumption of all cores does not exceed a predefined power constraint. Similarly, the work proposed in [7] adopts a chip-level power constraint and aims at maximizing the performance without exceeding the power constraint. The technique proposed in [6] allows exceeding the TDP constraint for a very short period of time, in order to activate some extra cores and thereby increase the performance of the multithreaded applications. To dissipate the additional heat generated during this period, a phase-change materiel is utilized in the thermal packaging. Nevertheless, these techniques that consider only a power constraint may still result in thermal violations (as shown in the first case study), which, in turn, lead to performance degradation.

Therefore, several techniques have been proposed to consider a temperature constraint while maximizing the performance. For example, the work in [25] optimizes the performance of an application under power and temperature constraints using DVFS and power-gating techniques. The work presented in [26] considers a temperature constraint while mapping the data flow graph (DFG) of an application. Another performance maximization technique is presented in [27] that maximizes the core speed by determining the v/f levels of the cores under a temperature constraint. However, the main limitation of this work is that it ignores the heat conductance among the cores in order to simplify the problem. More recent works have been proposed in [12, 13, 28] that consider dark silicon and aim at maximizing the performance and improving the thermal profile of the chip. However, these techniques do not employ DVFS; they rather consider one v/f level for all cores. In summary, none of these techniques that consider the temperature tackles the problem of distributing the cores among multiple applications.

Rather than increasing the performance, another branch of state-of-the-art techniques aims at increasing the dependability of the systems [29, 30] through managing the temperatures of the cores. For example, the approaches presented in [31, 32] attempt to reduce thermal violations on the chip, while scheduling the tasks of an application. The techniques proposed in [33, 34] target to minimize the peak temperature of the cores via employing multiagent systems. In [35], PID controllers are employed to jointly reduce the peak temperature and balance the temperatures throughout the chip. As it can be noticed, the state-of-the-art techniques under this branch do not distribute the chip's resources among multiple applications.

As a conclusion, there is a need for a dark silicon aware resource management technique that distributes the chip's resources among multiple applications and determines the v/f levels of the cores aiming at maximizing the overall system performance under a temperature constraint.

3. SYSTEM MODEL

DsRem targets a many-core chip that consists of N *homogeneous* cores with shared memory. This chip is a tiled architecture. Every tile has a group of cores and an L2 shared cache memory between its cores, where the L2 caches of the tiles are fully coherent. Every core has two power states: active (i.e., power-on) or dark (i.e., power-gated). Active cores can run at different speeds according to their v/f levels. A minimum voltage is necessary for a core to stably execute at a given frequency. In order to avoid consuming unnecessary power, the voltage of a core is always set to the minimum value such that stable execution can be achieved on the core for the desired frequency. These voltage and frequency pairs are enumerated as $\{vf_1, vf_2, \ldots, vf_{max}\}$. Table 1 summarizes all commonly used parameters in this chapter.

3.1 Application Model

The application model consists of K multithreaded applications, such that application threads can run in parallel on different cores. It is assumed that these applications are *malleable*, i.e., they can adapt the number of threads dynamically at runtime, based on the number of cores assigned to them [36]. TH_j denotes the number of threads of application j, which can range between 1 and TH_j^{max}, i.e., the maximum number of threads of application j.

To represent the mapping of applications to the cores, a binary matrix $\mathbf{M} = \left[m_{i,j} \right]_{N \times K}$ is defined. If application j is mapped to core i, $m_{i,j} = 1$, and $m_{i,\,j} = 0$ otherwise. It is assumed that only one application can be

Table 1 Commonly Used Parameters in This Chapter

Parameter	Description
N	Number of cores on the chip
Z	Number of the thermal nodes in the RC thermal network
K	Number of applications
$\mathbf{M} = \left[m_{i,j} \right]_{N \times K}$	Matrix of application-to-core assignment
TH_j	Selected number of threads of application j
TH_j^{max}	Maximum thread-level parallelism of application j
$V F_j$	Selected v/f level for the cores of application j
vf_{max}	Maximum v/f level available on the chip
D_j	Number of dark cores assigned to application j
PQ_j	The power quota assigned to application j
$\mathbf{P_j}$	Table of the total average power consumption (W) of application j for different number of cores and different v/f levels
$\mathbf{R_j}$	Table of the throughput of application j for different number of cores and different v/f levels
$\mathbf{B} = \left[b_{x,\gamma} \right]_{Z \times Z}$	$b_{x,\gamma}$ represents the heat contribution of node x into the temperature of node γ (kelvin/W)
T_{critical}	The critical temperature of the chip
$\mathbf{T}^{\text{core}} = \left[T_i^{\text{core}} \right]_{N \times 1}$	The steady-state temperatures of the cores

executed at one core, such that $\sum_{j=1}^{K} m_{i,j} \leq 1$ for all $i = 1, 2, ..., N$. Moreover, $\sum_{i=1}^{N} m_{i,j} = TH_j$, for all $j = 1, 2, ..., K$.

The total power consumption of each application is quantified according to the number of threads TH_j and the v/f level $V F_j$ that its cores run at. Therefore, a table $\mathbf{P_j}$ for each application is defined to store the average power consumption for all possible values of the number of threads, $th = 1, 2, ..., TH_j^{max}$, and the available v/f levels, $vf = vf_1, vf_2, ..., vf_{max}$. The $\mathbf{P_j}$ table is considered as an input to *DsRem* and it will be obtained by profiling the applications at design time. When an application thread is mapped to a core, the average power consumption on that core will be equal to $P_j(TH_j, V F_j)/TH_j$, assuming that the resulting average power consumptions on all application cores are equal. A column vector $\mathbf{P}^{App}$ is defined to store the average power consumptions for all applications as shown in Eq. (1). This table needs to be updated whenever the number of threads TH_j or the v/f level VF_j of any application are changed.

$$\mathbf{P}^{App} = \left[P_j(TH_j, VF_j) TH_j \right]_{K \times 1} \qquad (1)$$

Besides power, the application performance should be modeled. As a metric for performance, the inverse of the execution time of the application is used and denoted as the application throughput. For each application, a table $\mathbf{R_j}$ is defined to store the application throughput for all possible combinations of the number of threads, $th = 1, 2, \ldots, TH_j^{max}$, and the available v/f levels, $vf = vf_1, vf_2, \ldots, vf_{max}$. Each cell of the table $R_j(th, vf)$ contains the inverse of the corresponding execution time of the application $t_j(th, vf)$ as shown in Eq. (2). Similar to $\mathbf{P_j}$, $\mathbf{R_j}$ is also considered as an input to $DsRem$ and it will be obtained by profiling the applications at design time.

$$R_j(th, vf) = 1/t_j(th, vf) \qquad (2)$$

Note that, application throughput depends on the TLP and ILP of the application. When application j has more TLP than ILP, its throughput will be increased more by increasing the number of its threads TH_j. On the other hand, when application j has high ILP, its throughput will be increased more by increasing VF_j.

3.2 Thermal Model

For a thermal model, the well-known RC thermal network, defined in [37], is adopted. It considers the duality between thermal and electrical circuits. The RC thermal network consists of Z thermal nodes that represent the thermal blocks on the chip (shown in Fig. 10), the heat sink and the heat spreader, etc., such that $Z \geq N$. That implies, there are more thermal nodes in the thermal model than blocks on the floorplan. The thermal nodes are interconnected with each other through thermal conductances. The ambient temperature denoted as T_{amb}. The power consumptions of the cores and the other blocks on the chip are considered as heat sources. With these considerations, the temperature of every thermal node is a function of its power consumption, the temperatures of the neighboring nodes, and the ambient temperature.

According to this model, the steady-state temperatures on the thermal nodes can be calculated as follows:

$$\mathbf{T} = \mathbf{BP} + T_{amb}\mathbf{BG},$$

where the values in column vector $\mathbf{T}$ are the steady-state temperatures on the thermal nodes. The power consumptions of the thermal nodes are stored in column vector $\mathbf{P}$. Matrix $\mathbf{B}$ contains the amount of the heat contribution of the thermal nodes. Column vector $\mathbf{G}$ contains the heat contribution of the ambient temperature T_{amb} to the thermal nodes. Column vector $\mathbf{P}$ can be divided into two subvectors: $\mathbf{P}^{core}$ for the power consumptions on the cores; $\mathbf{P}^{uncore}$ for the power consumptions on the other blocks on the floorplan (e.g., L2 caches). Focusing on the thermal nodes that correspond to cores, the steady-state temperatures of the cores can be computed as follows:

$$\mathbf{T}^{core} = \mathbf{BP}^{core} + \mathbf{BP}^{uncore} + T_{amb}\mathbf{BG}, \tag{3}$$

where $\mathbf{T}^{core} = \left[T_i^{core} \right]_{N \times 1}$ are the steady-state temperatures of all N cores on the chip.

For simplicity, a column vector $\mathbf{C}$ is defined, such that it contains the last two terms of Eq. (3), i.e., the heat contribution of the other thermal nodes to the cores and the heat contribution of the ambient temperature T_{amb} to the cores. By replacing the last two terms of Eq. (3) with column vector $\mathbf{C}$, Eq. (4) is resulting.

$$\mathbf{T}^{core} = \mathbf{BP}^{core} + \mathbf{C}, \tag{4}$$

By compensating the power vector of the applications and the mapping matrix in Eq. (4), the following equation is resulting:

$$\mathbf{T}^{core} = \mathbf{BMP}^{App} + \mathbf{C}.$$

The following equation expresses the direct relation between the power consumptions of the applications and the steady-state temperature of any core i:

$$T_i^{core} = \sum_{j=1}^{K} \mathbf{B}_i \mathbf{M}_j \cdot \left(P_j(TH_j, VF_j) / TH_j \right) + C_i, \tag{5}$$

where $\mathbf{B}_i$ is the row i from matrix $\mathbf{B}$ which corresponds to core i. $\mathbf{M}_j$ is the column j of matrix $\mathbf{M}$ that corresponds to application j. Hence, the thermal model calculates the steady-state temperature for any core given the mapping matrix, besides to TH_j, $V F_j$ and the application power tables P_j for all applications.

4. PROBLEM DEFINITION

For a given chip composed of N cores and a set of K applications, the focus of *DsRem* is to select the number of threads for each application and the v/f level of the application cores, as well as the mapping of these applications to cores. The goal is to maximize the overall system throughput under the temperature constraint T_{critical}. Mathematically, this problem can be expressed as finding TH_j, $V F_j$, and matrix $\mathbf{M}_j$, for all $j = 1, 2, ..., K$, in order to:

$$\text{Maximize} \sum_{j=1}^{K} R_j(TH_j, VF_j)$$

$$\text{s.t.:} \sum_{j=1}^{K} TH_j \leq N \quad \text{and} \quad T_i^{\text{core}} < T_{\text{critical}} \quad \text{for all} \quad i = 1, 2, ..., N$$

Finding the optimal solution by exhaustive search through all the possible combinations of $\mathbf{M}_j$, TH_j, $V F_j$ for all applications requires exponential time complexity. Therefore, there is a need to decompose this problem into subproblems. The first subproblem is finding the TH_j and $V F_j$ for all applications, such that $\sum_{j=1}^{K} R_j(TH_j, VF_j)$ is maximized, where the total number of allocated cores should not exceed N. However, the solutions resulted from solving this problem could lead to many thermal violations, because the highest v/f levels would be chosen, as the goal is to maximize the throughput. Hence, in order to mitigate the potential thermal violations, TDP is considered as a chip-level power constraint in this subproblem. Thus, the two constraints of the first subproblem are as follows: $\sum_{j=1}^{K} TH_j \leq N$ and $\sum_{j=1}^{K} P_j(TH_j, VF_j) \leq TDP$. The second subproblem is determining the mapping of the applications to the available cores, i.e., $\mathbf{M}_j$, according to their allocated resources. The last subproblem is to modify the application resources TH_j, $V F_j$ and the mapping $\mathbf{M}_j$ by removing the TDP constraint and rather considering the temperature constraint T_{critical}, in order to avoid thermal violations and to exploit thermal headroom (if existing).

5. DARK SILICON AWARE RESOURCE MANAGEMENT

The presented dark silicon aware resource management technique, *DsRem*, manages the system resources and distributes them among multiple applications, by performing the following three steps that are corresponding to the aforementioned subproblems. At the first step, *DsRem* considers that TDP is the total power budget of the chip. Therefore, it optimally distributes both the cores and the TDP budget among the applications. To achieve this, *DsRem* employs a dynamic programming algorithm that jointly determines the number of the cores (threads) and the v/f levels of these cores for each application considering the TLP and ILP of the applications, such that the maximum overall system throughput is obtained. At the second step, the applications are mapped to cores according to their assigned resources in the first step. The goal of this step is to minimize the core temperatures, because this helps in reducing thermal violations on the chip. Therefore, before mapping the applications, the available dark cores are distributed among the applications in proportional to their power consumptions. Then, the applications are mapped such that the locations of active and dark cores of each application are close to each other, in order to allow the dark cores of each application to dissipate more heat from its active cores. The final step of *DsRem* considers the predefined temperature constraint and accordingly modifies the application settings obtained from the previous two steps in order to avoid thermal violations and exploit thermal headroom. In this step, TDP constraint is not considered and it might be exceeded, if there are available thermal headroom and thereby the resources are adapted to exploit it. Fig. 4 shows an overview of *DsRem* technique.

5.1 TDP-Constrained Optimal Resource Distribution

At the first step of *DsRem*, the system resources, i.e., the available cores and the power budget (TDP), will be distributed optimally among the applications in order to maximize the overall system throughput, which is the first subproblem defined in Section 4.

To solve this problem, *DsRem* employs an algorithm based on dynamic programming. Dynamic programming breaks down the problem into smaller subproblems and builds up the final solution gradually. The benefit here is that by storing the obtained subsolutions at every step in tracking

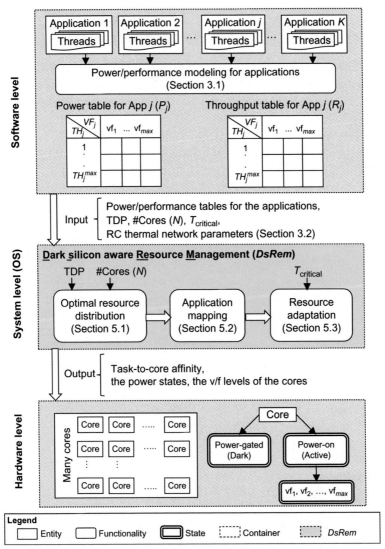

Fig. 4 Overview of the presented dark silicon aware resource management technique, *DsRem*.

tables, the algorithm can reuse them in further steps, thus reducing the number of combinations that should be tested. That, in turn, reduces the complexity and the required time for finding the final optimal solution.

The inputs of this algorithm are the application set with their power and throughput tables (see Section 3.1), besides the total number of cores N and the available power budget p_{max}, which is equal to TDP. The output will be the optimal settings TH_j, $V F_j$ for all $j = 1, 2, ..., K$ applications that result in the maximum overall system throughput. To achieve this goal, the algorithm needs to divide the total available power budget p_{max} to smaller power budgets (portions) p to be distributed among the applications similarly as n. The final determined power quota for application j is denoted as PQ_j.

Before explaining how this algorithm works, we need to define the following auxiliary function that the algorithm uses in finding the subsolutions. $MaxVF^{Budget}(j, n, p)$ returns the v/f level that results in the maximum throughput for application j under a given power budget p, where the number of threads is equal to n. If there is no v/f level available for the given power budget p, this function returns 0. An example of the work of this function is illustrated in Fig. 5. To find the optimal solution, the algorithm builds a table G^R, whose entries are represented by all possible unique triple (j, n, p), such that $j = 1, 2, ..., K$, $n = 1, 2, ..., N$, $p = p_1, p_2, ..., p_{max}$. Each cell of this table, $G^R(j, n, p)$, contains the maximum throughput for the first j applications (assuming an arbitrary order for the applications) using n cores and a peak power consumption less than or equal to p. It starts from the first application $j = 1$ and calculates its maximum throughput at all possible combinations between n and p, as shown in Eq. (6). This equation represents the initial boundary condition of building G^R table.

MaxVF ^{Budget}

Inputs: Power table for application j, number of threads TH = 6, power budget PQ = 15 W

P_j	VF$_1$ (2.0 GHz)	VF$_2$ (2.2 GHz)	VF$_3$ (2.4 GHz)	VF$_4$ (2.6 GHz)	VF$_5$ (2.8 GHz)
1	3.6	4.3	5.2	6.1	7.3
2	6.1	7.3	8.7	10.4	12.4
3	7.9	9.2	11.0	13.2	16.0
4	9.3	11.0	13.1	15.9	17.8
5	10.0	12.0	14.2	15.9	19.6
TH → 6	11.8	14.0	16.6	18.2	21.9
7	12.8	15.2	18.0	21.0	24.9
8	13.3	16.3	19.1	22.3	26.5

$MaxVF^{Budget}$ searches in the row of P_j that corresponds to TH, for the maximum power value that is less than PQ.

$P_j(6,1)$ $P_j(6,5)$

| 11.8 | 14.0 | 16.6 | 18.2 | 21.9 |

$P_j(6,5) > PQ$
$P_j(6,4) > PQ$
$P_j(6,3) > PQ$
$P_j(6,2) < PQ \rightarrow$ Return VF$_2$ (2.2 GHz)

Output: The maximum possible v/f level under the given power budget

Fig. 5 Example of the work of function $MaxVF^{Budget}(j, n, p)$.

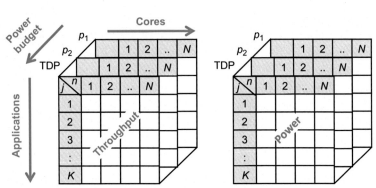

Fig. 6 The employed dynamic programming tables, $\mathbf{G^R}$ and $\mathbf{G^P}$, which have the same dimensions and, respectively, store the resulting throughput and power values for different combinations.

$$G^R(1,n,p) = \begin{cases} -\infty & \text{if } MaxVF^{Budget}(1,n,p) = 0 \\ R_j(n, MaxVF^{Budget}(1,n,p)) & \text{otherwise} \end{cases}$$

$$(6)$$

Besides $\mathbf{G^R}$, the algorithm builds an auxiliary table called $\mathbf{G^P}$ to store the resulting power consumption from obtaining the maximum throughput at each step. $\mathbf{G^P}$ has the same dimensions as $\mathbf{G^R}$ (see Fig. 6). The initial boundary condition of building $\mathbf{G^P}$ is shown in Eq. (7).

$$G^P(1,n,p) = \begin{cases} \infty & \text{if } MaxVF^{Budget}(1,n,p) = 0 \\ P_j(n, MaxVF^{Budget}(1,n,p)) & \text{otherwise} \end{cases}$$

$$(7)$$

To calculate $\mathbf{G^R}$ for $(j > 1)$ and $(j \leq n)$, the following recursive function is used:

$$G^R(j,n,p) = \max_{\substack{j-1 \leq n' < n \\ p_1 \leq p' < p}} \left\{ \begin{cases} -\infty & \text{if } G^P(j-1,n',p') + \\ & P_j(n-n', MaxVF^{Budget}(j,n-n',p-p')) > p \\ G^R(j-1,n',p') + & \\ R_j(n-n', MaxVF^{Budget}(j,n-n',p-p')) & \text{otherwise} \end{cases} \right\}$$

$$(8)$$

Using Eq. (8), the algorithm finds the maximum throughput out of all the possible divisions of n and p between the application j and the previous

$j - 1$ applications. To represent these divisions, the parameters n' and p' are used. n' will be set to the values from $j - 1$ to n, because at least $j - 1$ cores should be assigned to the previous $j - 1$ applications. We assume that $K \leq N$, and each application will get at least one. p' will be set to the values from $p1$ to p, covering the whole range of the power budgets, because there is no constraint on the minimum power budget that each application should get. The resulting throughput of all possible combinations of n' and p' will be calculated in order to choose the maximum one. As it can be noticed from Eq. (8), the previous computations of $G^R(j - 1, n', p')$ are reused in order to reduce the number of required calculations. When the maximum throughput out of these combinations is found, it is stored in $\mathbf{G^R}$, and the resulting power consumption from obtaining this throughput is stored in $\mathbf{G^P}$, as follows:

$$G^P \ (j, n, p) = \ G^P(j - 1, n', p') + P_j(n - n', MaxVF^{Budget}(j, n - n', p - p')),$$

(9)

where n' and p' are the values that result in the maximum throughput according to Eq. (8). In case there is no feasible solution for the given power budget p or there are more applications than n cores, $G^R(j, n, p)$ is set to $-\infty$ and $G^P(j, n, p)$ is set to ∞.

At the same time of building $\mathbf{G^R}$ and $\mathbf{G^P}$, the values of the parameters $n - n'$, $p - p'$, and $MaxV \ F^{Budget}(j, n - n', p - p')$, which leads to the maximum throughput that is stored in $G^R(j, n, p)$ at this step, are stored in tracking tables $TrackN$, $TrackP$, and $TrackVF$, respectively. The dimensions of these tables are the same as the dimensions of $\mathbf{G^R}$ and $\mathbf{G^P}$ tables. After building the tables, the final optimal solution, i.e., the maximum through-put, can be found in $\mathbf{G^R}$ at the cell that holds the maximum value. Finding this cell is performed by searching through the cells whose entries hold (K, p_{max}), to find the one that holds the maximum value. When this value is extracted, the corresponding n will be the total number of active cores n_{max} that the applications can run on in order to obtain the maximum throughput without exceeding the given power budget p_{max}, i.e., TDP. The optimal number of cores, the v/f level, and the power quota for each application will be derived from the tracking tables, starting from the entry (K, n_{max}, p_{max}) (which holds the maximum throughput). The values at this entry of the tracking tables $TrackN$, $TrackP$, and $TrackVF$ hold $n - n'$, $p - p'$, and $MaxV$

$F^{Budget}(j, n - n', p - p')$, respectively. These values represent the optimal settings $(TH_K, PQ_K, V F_K)$ of application K. For application $K - 1$, its optimal settings will be stored at the cell whose entry is $(K - 1, n', p')$, as it can be observed from Eq. (8). The values of n' and p' can be extracted by subtracting the found values of $n - n'$ and $p - p'$ from n and p, respectively. Similar process is repeated for further applications until reaching $j = 1$. The corresponding pseudo-code for building the tables $\mathbf{G^R}$ and $\mathbf{G^P}$ in a bottom-up manner is presented in Algorithm 1. The pseudo-code for extracting the solutions from the tracking tables is presented in Algorithm 2.

Algorithm 1 Build $\mathbf{G^R}$ and $\mathbf{G^P}$ Tables

```
Input: Application set with their power and throughput tables Pⱼ
and Rⱼ;
Output: Tables Gᴿ, Gᴾ, TrackN, TrackP, TrackV F;
 1: for j = 1 to K do
 2:    for n = 1 to N do
 3:       for p = p₁ to pₘₐₓ do
 4:          V F = MaxVFᴮᵘᵈᵍᵉᵗ(j, n, p);
 5:          if V F = 0 or j > n then
 6:             Gᴾ(j, n, p) ←∞;
 7:             Gᴿ(j, n, p) ←-∞;
 8:          else if VF > 0 and j ≤ n then
 9:             if j = 1 then
10:                Build Gᴿ according to Eq. (6);
11:                Build Gᴾ according to Eq. (7);
12:             else if Gᴾ(j - 1, n', p') + Pⱼ(n - n', MaxV Fᴮᵘᵈᵍᵉᵗ
                   (j, n - n', p - p')) ≤ p then
13:                Build Gᴿ according to Eq. (8);
14:                Build Gᴿ according to Eq. (9);
15:             end if
16:             Save the parameters n - n', p - p', MaxVFᴮᵘᵈᵍᵉᵗ
                   (j, n - n', p - p') that result in the maximum
                   throughput at this step in the tracking tables.
17:          end if
18:       end for
19:    end for
20: end for
21: return Tables Gᴿ, Gᴾ, TrackN, TrackP, TrackVF;
```

Algorithm 2 Extract Optimal Application Settings TH_j, PQ_j, $V F_j$, From Dynamic Programming Tracking Tables

```
Input: Tables GR, GP, TrackN, TrackP, TrackV F;
    {Find the total number of active cores}
 1: nmax = 0;
 2: MaxGR = 0
 3: for n = N to 1 do
 4:    if GR(K, N, pmax) > MaxGR then
 5:       MaxGR = GR(K, N, pmax);
 6:       nmax = n;
 7:    end if
 8: end for
    {Find the optimal settings of applications}
 9: n = nmax;
10: p = pmax;
11: for j = K to 1 do
12:    THj = Track_N(j, n, p);
13:    PQj = Track_P(j, n, p);
14:    V Fj = Track_V H(j, n, p);
       {The entries for the optimal settings of the previous j - 1
       applications}
15:    n - n - Track_N(j, n, p);
16:    p = p - Track_P(j, n, p);
17: end for
18: return THj, PQj, V Fj for all applications.
```

Hence, by the end of this step, the optimal settings of each application j are determined, namely, the number of threads (its active cores), TH_j, the v/f level of its cores $V F_j$, and its power quota PQ_j.

Proof of optimality: We prove the optimality of the presented resource distribution algorithm by induction. Initially, $\forall n$, p, $G^R(1, n, p)$ is optimal because all the values of $R_1(n, p)$ and $P_1(n, p)$ are tested, and the maximum throughput under the give power budget p is stored in $G^R(1, n, p)$. Now we need to prove that $G^R(k, n, p)$ is optimal assuming that $G^R(k - 1, n, p)$ is optimal.

To achieve this, we first assume that Eq. (8) does not result in the maximum throughput for $G^R(k, n, p)$. According to Eq. (8), the value of $G^R(k, n, p)$ is obtained by calculating the resulting throughput for all possible combinations of dividing n and p between the previous $k - 1$ applications and the current k application.

Thus, for each combination, different portions n', p' are given to the previous $k - 1$ applications while $n'' = n - n'$ and $p'' = p - p'$ are given to the current application k. $\forall n, p$, the term $R_k(n'', maxV F^{Budget}(k, n'', p''))$ gives the throughput of the application k at n'' cores and at v/f level equal to $maxV$ $F^{Budget}(k, n'', p'')$. The function $maxV F^{Budget}(k, n'', p'')$ returns the maximum v/f level that leads to a power consumption less than p'' where the number of cores is n''. By definition the maximum v/f level (for n'' and p'') will give the maximum throughput of application k (for n'' and p''). As a result, the term $R_k(n'', maxV F(n'', p''))$ will give the optimal solution for application k for a specific n'', p''. Consequently, $\exists n', p'$ that make the term $G^R(k - 1, n', p')$ not optimal, and that, however, contradicts the assumption. As a result, by the mathematical induction hypothesis, we reach the conclusion that $G^R(k, n, p)$ is optimal $\forall n, p$.

5.2 Thermal-Aware Application Mapping

After determining the optimal number of TH_j and $V F_j$ of each application j in the previous step, $DsRem$ employs a heuristic for application mapping that specifies which cores the applications will be mapped to. Since the previous dynamic programming algorithm considers the TDP constraint, some cores might be left dark and not assigned to any application. In this step, $DsRem$ aims to exploit dark silicon to dissipate the generated heat on the active cores. Therefore, $DsRem$ will first distribute the dark cores among the applications proportional to their power consumptions. Thus, the applications with higher power consumptions get more dark cores than the applications with lower power consumptions. The amount of dark cores specified for application j is denoted as D_j. After determining the number of dark cores of each application, the application mapping policy tries to specify close locations for the active and dark cores of each application, in order to allow the dark cores of each application to dissipate some of the generated heat on its active cores. Moreover, the worst fit decreasing (WFD) algorithm is employed within the application mapping policy in order to map the applications with high power consumption to tiles with low average temperature. That also helps in reducing the resulting temperatures of the cores.

Thus, the application mapping policy of $DsRem$ consists of the following steps: (1) Find the unmapped application j that has the highest power consumption. (2) Find the coldest tile L, i.e., the average of its core temperatures is the lowest compared to the other tiles. (3) If tile L does not have $TH_j + D_j$ available cores, the policy reserves only the available cores of L

and searches for available cores in neighboring tiles until the whole TH_j+D_j cores are reserved. Hence, a subset of cores are reserved for application j, namely, $listCores_j$. Afterward, the policy (4) chooses the coldest core of $listCores_j$, then (5) maps one thread to it and reestimates the core temperatures. The last two steps are repeated until all application threads are mapped. As a result of this policy, the active cores are distributed loosely through the region of the application cores $listCores_j$ and the dark cores will be surrounding them. Moreover, the region of cores that will be assigned to the next application will be automatically far from the first application region, because the presented policy will reestimate the core temperatures after mapping each application and then searches for the coldest tile. The whole steps are repeated until mapping all applications. Fig. 7 illustrates an example about mapping a list of applications to the chip.

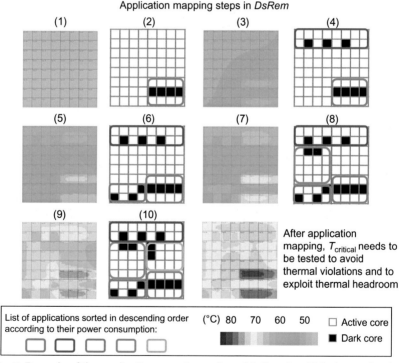

Fig. 7 Example of the application mapping policy of *DsRem*, in which the applications are iteratively mapped to the chip. At each iteration, the core temperatures will be estimated, in order to map the application with the highest power consumption to the coldest region on the chip.

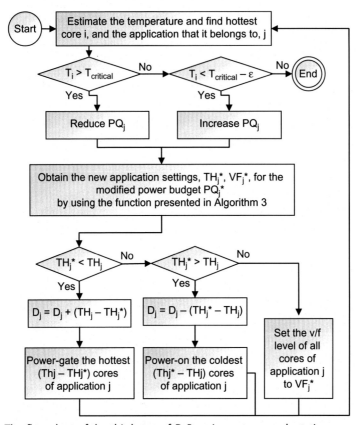

Fig. 8 The flow chart of the third step of *DsRem*, i.e., resource adaptation.

5.3 Thermal-Constrained Resource Adaptation

Although the presented mapping policy in the previous section aims at minimizing the potential temperature increase, it is still not enough to avoid thermal violations. Therefore, a heuristic to adapt the resources considering the temperature constraint is proposed in order to avoid thermal violations and to exploit available thermal headroom. The flow chart of the proposed heuristic is shown in Fig. 8. The first step of this heuristic is to compute the core temperatures using Eq. (5) considering the application settings and mapping determined in the previous steps. In case, any core of application j exceeds $T_{critical}$, the power quota PQ_j that is assigned to that application at step 1 (see Section 5.1) should be decreased. Contrarily, if there is available thermal headroom on some of the cores of an application, the assigned power quota of the application PQ_j will be increased to ultimately increase the throughput. Note that *DsRem* does not increase the power quota of the application, if the peak temperatures of the application cores is slightly below the critical temperature, because this increase might lead to exceed T_{critical}, and thereby PQ_j must be reduced in the next iteration of the algorithm leading to unstable behavior. To prevent this, *DsRem* employs a parameter ε to determine if there is a *sufficient* thermal headroom to increase the power quota or not. This value needs to be empirically determined.

After increasing or decreasing the power quota PQ_j, new application settings TH_j, $V F_j$ should be obtained, such that the throughput of application j is maximized under the new power quota. Finding these new settings can be done by searching within the throughput and power tables of that application, as shown in Algorithm 3. Note that the new TH_j can be increased up to $\min\left\{TH_j^{max}, TH_j + D_j\right\}$. In other words, the application will adapt its own resources for maximizing its throughput under the new power quota. Note that, if this algorithm leads to increase TH_j that means dark cores of application j need to be activated to run threads. In this case, the coldest dark cores are activated. Contrarily, if TH_j is decreased, the hottest active cores of application j are darken.

Algorithm 3 Find the Settings of Application j That Maximize Its Throughput Under a Power Budget p

```
Input: Power and performance tables Pj and Rj (Section 3.1);
Output: TH, V F;
 1: MaxR = 0; TH* = 0; V F* = 0;
 2: for th = Min(TH_j^max, TH_j + D_j) to 1 do
 3:    for vf = vf_max to vf_1 do
 4:       if P_j(th, vf) < p and R_j(th, vf) > MaxR then
 5:          MaxR = R_j(th, vf);
 6:          TH* = th;
 7:          V F* = vf;
 8:       end if
 9:    end for
10: end for
11: return TH*, V F*;
```

After obtaining the new application settings, the steady-state temperatures of the cores will be reestimated again. If there are still some cores that exceed T_{critical}, PQ_j will be decreased again. If PQ_j is decreased to a value, at which not even one thread at the lowest v/f level can run without exceeding T_{critical}, TH_j is set to zero. In other words, all the cores of that application become dark. That means, running application j under the temperature constraint $T_{critical}$ is not feasible. Fig. 8 shows the flow chart of the resource adaptation heuristic.

Applying the first and the second steps of *DsRem* results in a thermal-aware application mapping with optimal resource distribution under TDP constraint. We refer to this combination as *TDPmap*. Nevertheless, having a TDP-constrained solution might still lead to thermal violations or to thermal headroom. Thus, the third step (resource adaptation) is applied and it might change the optimal resource distribution obtained by the first step, in order to consider the temperature constraint. Considering the temperature constraint, *DsRem* at the third step can adapt the resources in order to avoid thermal violations and to exploit available thermal headroom, while taking into account the performance and the power models of the applications, aiming at maximizing the overall system throughput under the temperature constraint. An overview of the three steps of *DsRem* is shown in Fig. 9.

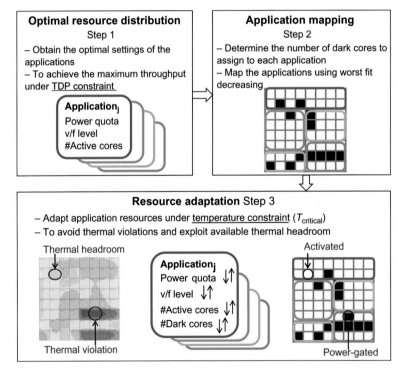

Fig. 9 The steps of the presented dark silicon resource management technique, *DsRem*.

6. EXPERIMENTAL EVALUATIONS

6.1 Setup

To evaluate *DsRem*, a simulated 64-core chip is employed. This chip is tiled architecture similar to Intel's Single Chip Cloud computer (SCC) [38], but with more cores and more tiles. The number of the tiles in the system is 8. Each tile has 8 cores and one shared L2 cache of 2 MB, and a memory controller. Each core is an out-of-order Alpha 21264, simulated by gem5 [14], where Alpha 21264 represents an advanced real-world processor architecture [39]. The areas of the cores and the other components are estimated using McPAT [1], where the technology node is set to 22 nm. The floorplan of the chip consists of 80 thermal blocks, one for each core, and the rest of the blocks correspond to the L2 caches and the memory controllers. Fig. 10 shows the adopted floorplan, with its dimensions. According to the conducted simulations, each core has an area of 9.6 mm^2. The combined area of the L2 cache and the memory controller of each tile is 9.4 mm^2.

HotSpot [37] is used to obtain the parameters of the RC thermal network for the adopted floorplan. These parameters, i.e., **B**, **G**, are employed in calculating the temperatures of the cores as explained in the adopted thermal model (see Section 3.2). The default configurations of HotSpot are used: chip thickness of 0.15 mm, silicon thermal conductivity of $100\frac{W}{m \cdot K}$, silicon specific heat of $1.75 \cdot 10^6 \frac{J}{m^3 \cdot K}$, a heat sink of 6×6 cm and 6.9 mm thick, heat sink convection capacitance of $140.4\frac{J}{K}$, heat sink convection resistance

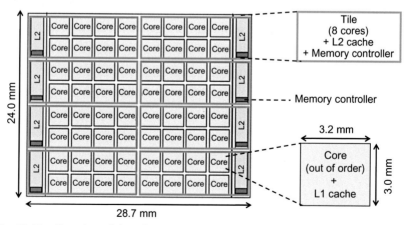

Fig. 10 The floorplan of the adopted 64-core chip.

of $0.1\frac{K}{W}$, heat sink and heat spreader thermal conductivity of $400\frac{W}{m \cdot K}$, heat sink and heat spreader specific heat of $3.55 \cdot 10^6 \frac{J}{m^3 \cdot K}$, a heat spreader of 3×3 cm and 1 mm thick, interface material thickness of 20 um, interface material thermal conductivity of $4\frac{W}{m \cdot K}$, and interface material specific heat of $4 \cdot 10^6 \frac{J}{m^3 \cdot K}$. The critical temperature, $T_{critical}$, is set to 80°C. The ambient temperature T_{amb} is 45°C. The thermal design power (TDP) is 220W, which is near the practical numbers used in a real processor, i.e., Intel Xeon Phi [40], that consists of 64 cores.

For applications, x264, ferret, blackscholes, and bodytrack from PAR-SEC benchmark suite [15] are utilized. Some of them have TLP more than ILP, like blackscholes, and some have ILP more than TLP, like ferret. Thus, these benchmarks are suitable examples to evaluate the efficiency of *DsRem*. As shown in Table 2, five scenarios (mix1, mix2, mix3, mix4, mix5) of multiple instances of these applications are composed in order to provide various workloads with different ILP and TLP. Using gem5, each application is executed under different number of threads $\{1, 2, \ldots, 8\}$ and different v/f levels. The range of the frequencies is $\{0.2, 0.4, \ldots, 4.0\}$ GHz, and the corresponding voltage settings for these frequencies are set according to [41]. The simulator gem5 generates performance traces of the applications, which are used by McPAT to estimate the resulting power consumptions. Using the generated power and performance traces by gem5 and McPAT, the power and throughput tables of the applications are obtained. Note that, cache and memory accesses of the threads of each application are simulated by gem5, which will be implicitly reflected in the power and throughput tables. The complete tool flow of the experimental setup is illustrated in Fig. 11.

It is assumed that *DsRem* will be implemented at design time, because deriving the optimal solution at step 1 requires a considerable time (2.36s).

Table 2 The Adopted Scenarios of Different Applications

Scenario	Blackscholes	x264	Bodytrack	Ferret
Mix1	1	2	2	3
Mix2	—	3	1	4
Mix3	—	4	—	4
Mix4	4	—	—	4
Mix5	4	—	4	—

Each scenario consists of multiple instances of different applications. Each cell represents the number of instances of each application in the corresponding scenario.

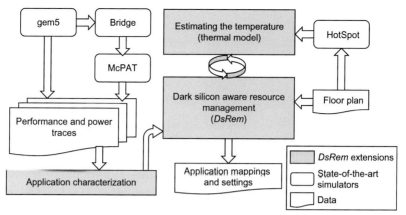

Fig. 11 The flow of the experimental setup, in which three state-of-the-art simulators, i.e., gem5 [14], McPAT [1], HotSpot [42], have been used together with the presented technique.

However, runtime evaluation is performed as well to check the resulting transient temperatures, when the decisions of *DsRem* w.r.t. the application resources and their mappings are given as inputs to runtime simulation. In the following subsections we present various results of comparing *DsRem* with different state-of-the-art techniques.

6.2 Results

Before evaluating *DsRem* in general and comparing it to state-of-the-art techniques, it is important to evaluate its individual steps. The first step of *DsRem* results in the optimal solution; therefore, there is no need to evaluate it or compare it with other techniques. However, the second and the third steps of *DsRem* are heuristics. Therefore, they are evaluated in the following first two sections.

6.2.1 Evaluation of the Presented Mapping Policy (the Second Step of DsRem)

As aforementioned, at the second step of *DsRem*, i.e., application mapping, a heuristic based on WFD is employed and aims at reducing the temperature while mapping the applications. To evaluate this heuristic, it is compared to another application mapping policy, referred to as *ContiguousMap*, which tries to keep both the cores, that applications are mapped to, and the idle cores contiguous. Several application mapping techniques, e.g., [21, 23], adopt the concept of contiguous mapping and map the applications close to each other in order to reduce the area fragmentation of the chip.

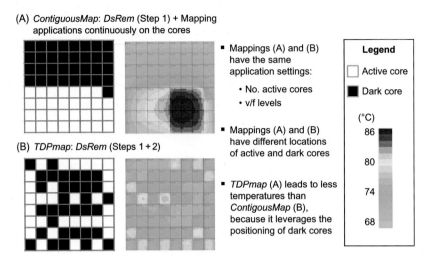

(A) *ContiguousMap*: DsRem (Step 1) + Mapping
applications continuously on the cores

- Mappings (A) and (B) have the same application settings:
  - No. active cores
  - v/f levels

(B) *TDPmap*: DsRem (Steps 1 + 2)

- Mappings (A) and (B) have different locations of active and dark cores

- *TDPmap* (A) leads to less temperatures than *ContigousMap* (B), because it leverages the positioning of dark cores

Legend

☐ Active core

■ Dark core

(°C)
86
80
74
68

Fig. 12 Comparison between the presented mapping policy, *TDPmap*, i.e., the second step of *DsRem*, and a contiguous mapping policy. Both policies utilize the same optimal application settings obtained by the first step of *DsRem*.

For fair comparison, the results of the first step of *DsRem*, i.e., the optimal application settings, are provided as inputs to *ContiguousMap*. As aforementioned, the version of *DsRem* that only executes the first and second steps is referred to as *TDPmap*. Hence, the comparison is conducted between *TDPmap* and *ContiguousMap*. Both of these techniques consider the same application settings; the number of active cores (threads) of the applications, and the v/f levels of their cores, while each of them applies its own mapping policy. We compare the resulting thermal maps of applying the *ContiguousMap* and *TDPmap*, as seen in Fig. 12. As it can be observed, the incurred temperatures on the cores when applying the mapping policy *TDPmap* are less than the incurred ones when applying *ContiguousMap*. Decreasing the core temperatures helps in reducing the potential thermal violations and increasing the available thermal headroom on the cores. That facilitates achieving the goal of the third step of *DsRem*, i.e., avoiding thermal violations and exploiting thermal headroom.

6.2.2 Evaluation of the Presented Resource Adaptation Policy (the Third Step of DsRem)

The focus of this section is to evaluate the importance of the last step of *DsRem*, i.e., resource adaptation. For this purpose, several experiments for various workload scenarios are conducted to evaluate the resulting

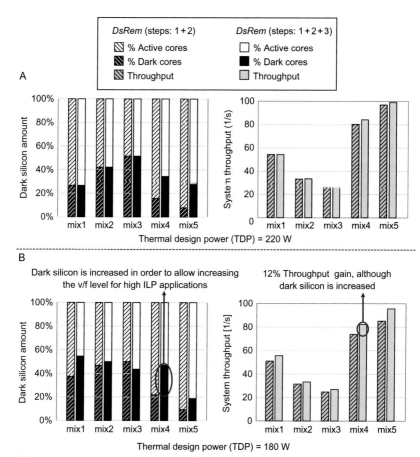

Fig. 13 Evaluation of the presented resource adaptation, i.e., the third step of *DsRem*, by comparing *DsRem* with *TDPmap*, i.e., the combination of the first and the second steps of *DsRem*.

system throughput when applying *DsRem* without resource adaptation, which means applying only the first and the second step of *DsRem*, i.e., *TDPmap*. Then, we compare the resulting throughput of *TDPmap* with the resulting one when applying *DsRem* with all its three steps. Besides the system throughput, the resulting amount of dark silicon is evaluated as well. Moreover, these experiments are conducted for two TDP values, i.e., 220 and 180 W. The resulting throughput of these experiments is shown in Fig. 13.

The first observation of this figure is that, the differences between the results of *DsRem* and *TDPmap* (w.r.t. the amount of dark silicon and the system throughput) are more at the lower value of TDP. The reason is that, setting TDP to a low value leads to less temperatures on the cores, and potentially larger thermal headroom exists on the cores. Hence, the last step of *DsRem* would have more possibilities to adapt the resources such that the available thermal headroom is exploited. The second observation is that, the resource adaptation step either increases dark silicon amount or decreases it according to the TLP and ILP of the applications. For instance, at mix3 of scenario (b), the resource adaptation decreases the amount of dark silicon while the throughput is increased. That means, there was available thermal headroom at some cores, and it is exploited by increasing the number of threads of one or more applications from the running workload. On the other hand, at mix4 the amount of dark silicon is increased by the resource adaptation. Nevertheless, the throughput is increased as well. Since the throughput is increased that means the adaptation in this case is also to exploit available thermal headroom, and not to avoid thermal violations. As a result, the resource adaptation in this case increases the throughput by increasing the amount of dark cores of one or more applications from the workload and at the same time increasing the v/f levels of their active cores. In summary, the gain in the system throughput resulting by applying the resource adaptation step is 12% compared to *TDPmap*.

6.2.3 Comparison Between DsRem *and a State-of-the-Art Technique in Throughput Maximization*

To achieve this comparison, the DVFS policy presented in [27] is implemented and denoted as *MaxSpeed*. *MaxSpeed* aims at maximizing the system throughput under the temperature constraint. Hence, *DsRem* and *MaxSpeed* share the same goal. However, the authors in *MaxSpeed* assume that the lateral heat conductance between cores is neglected, because their adopted floorplan contains large caches around cores similar to a chessboard pattern. Therefore, to enable comparison with *MaxSpeed*, the chessboard pattern is adopted for the active and dark cores, as shown in Fig. 14, *MaxSpeed* is applied on the active cores. Furthermore, *MaxSpeed* does not target multithreaded applications and thereby does not determine the number of cores to assign to each application. Therefore, it is assumed that the active cores will be equally distributed to the coming applications in the case of *MaxSpeed*. Then, *MaxSpeed* determines the v/f levels of the active cores, such that the throughput is maximized while the core temperatures are kept below T_{critical}, regarding their adopted thermal model that neglects the lateral heat transfer. *MaxSpeed* is applied on all scenarios (mix1, ..., mix5) of different applications. For example, in Fig. 14, the resulting thermal maps

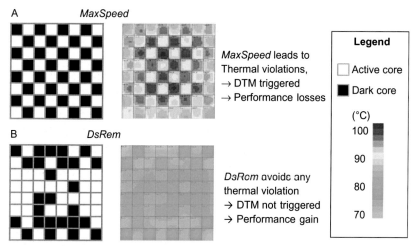

Fig. 14 Comparison between *DsRem* and *MaxSpeed*, where the latter leads to thermal violations due to its ignorance of the lateral heat transfer between the cores.

is shown after applying *MaxSpeed* and *DsRem* on mix5. Note that the thermal maps are calculated using the adopted thermal model (Section 3.2), in which the lateral heat transfer between the cores are considered. As it can be noticed from the shown thermal maps, *MaxSpeed* leads to a temperature increase more than 100°C, due to the ignorance of the heat contribution of the neighboring cores. This experiment demonstrates the importance of considering the heat transfer among the cores in the employed thermal model to estimate the temperatures more accurately.

In general, when the decisions of a resource management technique lead to thermal violations, DTM on the chip will be triggered to throttle down the cores and save the chip from burning out. Therefore, a simple thermal management technique is implemented in this experiment to check the core temperatures after the resource allocation decisions. If there is any thermal violation, the employed thermal management throttles down the v/f levels to keep the core temperatures below T_{critical}. Since *MaxSpeed* leads to thermal violations, the employed thermal management will throttling down the v/f levels, and consequently the system throughput will be decreased. Contrarily, *DsRem* does not lead to any thermal violation. Fig. 15 shows the resulting throughput of *DsRem* compared to the resulting one of the *MaxSpeed* after throttling down the v/f levels by the employed thermal management. As shown in the figure, the throughput gain of *DsRem* compared to *MaxSpeed* reaches up to 46%. On average, the throughput gain of *DsRem* is 34%.

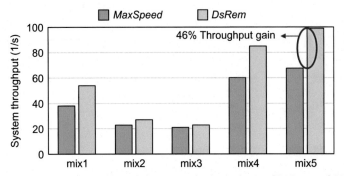

Fig. 15 Comparison between the resulting system throughput of *DsRem* and *MaxSpeed* [27]. The throughput gain of *DsRem* is up to 46% compared to *MaxSpeed*.

6.2.4 Comparison Between DsRem *and the State-of-the-Art Boosting Technique*

As noticed, the first comparison candidate results in severe thermal violations due to its employed thermal model that ignores the heat transfer among the cores. Therefore, we adopt another comparison candidate, which is a boosting technique that determines the v/f level of the cores based on measuring the temperatures of the cores at runtime within a control loop, similar to Intel Turbo Boost [43]. This boosting technique, referred to as *Boosting*, shares the same goal of *DsRem*, i.e., maximizing the performance under the temperature constraint. To achieve its goal, *Boosting* checks the peak temperature of the cores at each control period; if the peak temperature is below $T_{critical}$, *Boosting* increases the v/f levels of the cores by one step. Otherwise, if the peak temperature is higher than $T_{critical}$, the v/f levels are decreased by one step. This process will be repeated over the execution time. Hence, *Boosting* exploits any available thermal headroom at runtime.

However, *Boosting* does not determine the number of active and dark cores. Moreover, it does not determine the number of the cores that should be allocated to each application. Therefore, we adopt the chessboard pattern for the active and dark cores and consider that the active cores will be equally distributed to the coming applications. After that, *Boosting* will boost the v/f levels of the cores at runtime. Therefore, a simulation loop is executed for 500 ms, and *Boosting* is activated to take its decision at each control period (5 ms). To compare *DsRem* with *Boosting*, *DsRem* is applied at design time to determine the application mapping with their settings. Then, a simulation loop is executed for 500 ms, and at each control period (5 ms), the transient

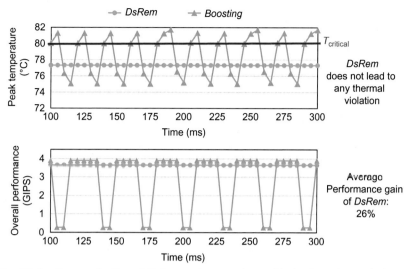

Fig. 16 Runtime evaluation of *DsRem* and comparison with the boosting technique.

temperatures of the cores are checked. If *DsRem* leads to thermal violations, the DTM technique will be triggered to throttle down the v/f level in order to reduce the core temperatures.

In Fig. 16, we show first the resulting peak temperature over execution time when applying *DsRem* and *Boosting*. We can observe that, *DsRem* does not lead to any thermal violation and thereby DTM has not been triggered. However, *Boosting* leads to increase in the peak temperature (due to the increase in the v/f levels) until exceeding $T_{critical}$. Once exceeded, *Boosting* throttles down the v/f levels. This process will be repeated over the simulation time. The second curve of Fig. 16 shows the resulting instructions per second (IPS) of the running workload in the case of *DsRem* and *Boosting*. As noticed, *Boosting* can lead to higher IPS at some points of the execution time, but on average, *DsRem* has 26% higher IPS than *Boosting*.

As a summary of the evaluation, *DsRem* efficiently manages the resources of the chip considering dark silicon. The performance gain that *DsRem* achieved is up to 46%, and an average of 34%. Moreover, the evaluation shows that maximizing the overall performance over time (*DsRem*) results in better performance gain on average, compared to maximizing the performance during short-time intervals (*Boosting*).

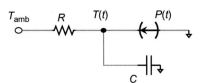

Fig. 17 Simple RC thermal network example for one core directly in contact with the ambient temperature.

6.3 Evaluation of the Temporal Thermal Gradient

It can be observed from Fig. 16 that the transient temperature increases by about 5°C after each control period (5ms). In this section, we present a theoretical analysis to evaluate the observed temporal thermal gradient in the simulations, similar to the analysis conducted in [44].

In this analysis, a simple RC thermal network is considered as shown in Fig. 17. The adopted RC thermal network consists of one thermal node. This node represents only the silicon layer of one core that is directly in contact with the ambient temperature. The cooling system and the other layers of the core have not been considered in this theoretical analysis to simplify the calculations. This assumed simple RC thermal network does not consider all the factors that existed in the complete RC thermal network that represents the whole chip. These factors, in turn, can influence the temperature increase on the cores. An example of these factors is the heat conductance between the cores. Therefore, the goal of this theoretical example is to evaluate the order of magnitude of the observed thermal gradient.

From Fig. 17 and based on Kirchhoff's first law, the temperature equation can be derived as follows:

$$P(t) = \frac{T(t) - T_{amb}}{R} + C\frac{dT(t)}{dt}$$

$P(t)$ is a step function, such that $P(t) = P_x$ at $t = 0$. Thus, we can write $P(t)$ using the Heaviside function $\Theta(t)$ as follows:

$$P_x\Theta(t) - \frac{T(t) - T_{amb}}{R} - C\frac{dT(t)}{dt}$$

This equation can be solved using *Laplace Transform*, as follows:

$$\frac{P_x}{s} - \frac{T(s)}{R} + \frac{T_{amb}}{sR} - CsT(s) + CT_{amb} = 0$$

Rearranging this equation, the following equation is resulting:

$$T(s) = \frac{\dfrac{P_x}{s} + \dfrac{T_{amb}}{sR} + CT_{amb}}{\dfrac{1}{R} + Cs}$$

By applying *Inverse Laplace Transform*, the transient temperature of the core $T(t)$ can be calculated as:

$$T(t) = -P_x Re^{-\frac{t}{R \cdot C}} + P_x R + T_{amb}$$

The derivative $T'(t)$ can be computed as:

$$T'(t) = \frac{P_x}{C} e^{-\frac{t}{R \cdot C}}$$

The maximum value of the derivative which represents the maximum temporal thermal gradient can be obtained at $t = 0$:

$$T'_{max}(t) = \frac{P_x}{C}$$

In the experiment shown in Fig. 16, the *Boosting* technique increases the v/f levels of the cores from 3.4 to 4 GHz. This increase of the v/f levels causes 4W increase in the power consumption of the cores, and thereby leading to that observed temperature increase. Therefore, we use the same power value for the theoretical analysis and set $P_x = 4$. To calculate C and R, we use the same configurations adopted in the simulations as follows.

We assume a chip thickness H of 0.15 mm, a silicon thermal conductivity k of $100 \frac{W}{m \cdot K}$, a silicon specific heat S of $1.75 \cdot 10^6 \frac{J}{m^3 \cdot K}$, and a core area A from McPAT of 9.6 mm^2. Then, the values of R and C are

$$R = \frac{H}{k \cdot A} = 156.25 \cdot 10^{-3} \frac{K}{W}, \quad C = S \cdot H \cdot A = 2.52 \cdot 10^{-3} \frac{W\,s}{K}$$

Given the adopted configurations, $T(t)$ and $T'(t)$ are calculated and plotted in Fig. 18. As a result, the maximum thermal gradient on silicon for a power increase of 4 W is equal to 1.6°C/ms. For 5ms, the maximum temperature increase is 7.9°C. Hence, the observed thermal gradient by the conducted simulations satisfies the theoretical limit of thermal gradient.

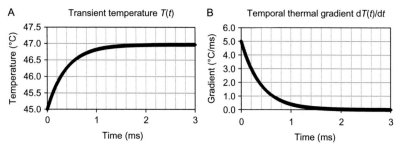

Fig. 18 The resulting transient temperature and its derivative in the simple RC thermal network shown in Fig. 17.

7. DARK SILICON AWARE RESOURCE MANAGEMENT FOR HETEROGENEOUS MANY-CORE SYSTEMS

As aforementioned, *DsRem* is proposed to manage the resources of homogenous many-core systems. For heterogeneous ones, the problem of managing the resources considering dark silicon is more complicated. The reason is that, the power and performance characteristics of applications significantly differ from core type to another in heterogeneous systems. Thus, resource management techniques should take these different characteristics into account while making their decisions. Furthermore, the thermal characteristics of heterogeneous many-core systems are also different. Even if two heterogeneous cores have the same power value, the resulting temperatures, the two exhibit, might be different if they have different areas. Consequently, the power and thermal constraints adopted in homogenous many-core systems cannot be simply employed in heterogeneous ones. Therefore, a novel power density constraint is derived in [44] that guarantees avoiding thermal violations and considers core heterogeneity within a chip and heat transfer among cores. Hence, the derived power density constraint enables resource management techniques from making their decisions of application mapping abstracted from thermal issues.

Based on the derived power density constraint, a resource management technique, referred to as *PdRM* [44], has been proposed for heterogeneous many-core systems. *PdRM* maps the applications to the suitable types of cores and determines their v/f levels, aiming at maximizing the overall system performance under the derived power density constraint. After mapping the applications, *PdRM* can adapt the power density constraint in order to react to workload changes at runtime and to exploit any available thermal headroom. During adaptation, *PdRM* considers the power characteristics of the executed applications on the heterogeneous cores. The evaluation of *PdRM* shows its superiority compared to several state-of-the-art techniques.

8. CONCLUSIONS

Although dark silicon phenomenon can be considered as a problem, it can also be leveraged by resource management techniques in order to deliver more performance. In this chapter, we have discussed a resource management technique, called *DsRem*, that leverages the positioning of dark silicon to dissipate more heat from the active cores, which ultimately allows increasing the v/f levels of the active cores, and thereby improving the overall system performance.

Considering dark silicon, *DsRem* efficiently distributes the resources among applications by jointly deciding the number of the cores that should be assigned to each application and the v/f level of these cores, such that the overall system performance is maximized. The temperature constraint of the chip is considered as an essential factor by the decision-making process of *DsRem*. As the conducted evaluation shows, the decisions of *DsRem* do not lead to any thermal violations and it exploits available thermal headroom in order to maximize the performance.

ACKNOWLEDGMENTS

This work was partly supported by the German Research Foundation (DFG) as part of the Transregional Collaborative Research Centre *Invasive Computing* (SFB/TR 89: http://invasic.de).

REFERENCES

[1] S. Li, J.-H. Ahn, R.D. Strong, J.B. Brockman, D.M. Tullsen, N.P. Jouppi, McPAT: an integrated power, area, and timing modeling framework for multicore and manycore architectures, in: 42nd IEEE/ACM International Symposium on Microarchitecture (MICRO), 2009, pp. 469–480.
[2] International Technology Roadmap for Semiconductors (ITRS), http://www.itrs.net.
[3] R. Borkar, M. Bohr, S. Jourdan, Advancing Moore's Law in 2014, https://www.intel.com/content/www/us/en/silicon-innovations/advancing-moores-law-in-2014-presentation.html.
[4] H. Esmaeilzadeh, E. Blem, R.St. Amant, K. Sankaralingam, D. Burger, Dark silicon and the end of multicore scaling, in: The 38th International Symposium on Computer Architecture (ISCA), 2011, pp. 365–376.
[5] J. Henkel, H. Khdr, S. Pagani, M. Shafique, New trends in dark silicon, in: 52nd Annual Design Automation Conference (DAC), 2015.

[6] A. Raghavan, Y. Luo, A. Chandawalla, M. Papaefthymiou, K.P. Pipe, T.F. Wenisch, M.M.K. Martin, Computational sprinting, in: 2012 IEEE 18th International Symposium on High Performance Computer Architecture (HPCA), IEEE, 2012, pp. 1–12.

[7] T.S. Muthukaruppan, M. Pricopi, V. Venkataramani, T. Mitra, S. Vishin, Hierarchical power management for asymmetric multi-core in dark silicon era, in: 50th Design Automation Conference (DAC), 2013, pp. 174:1–174:9.

[8] B. Raghunathan, Y. Turakhia, S. Garg, D. Marculescu, Cherry-picking: exploiting process variations in dark-silicon homogeneous chip multi-processors, in: DATE, 2013, pp. 39–44.

[9] A. Rezaei, D. Zhao, M. Daneshtalab, H. Wu, Shift sprinting: fine-grained temperature-aware NoC-based MCSoC architecture in dark silicon age. in: 2016 53nd ACM/EDAC/IEEE Design Automation Conference (DAC), 2016, pp. 1–6, https://doi.org/10.1145/2897937.2898090.

[10] S. Pagani, H. Khdr, W. Munawar, J.-J. Chen, M. Shafique, M. Li, J. Henkel, TSP: thermal safe power - efficient power budgeting for many-core systems in dark silicon, in: International Conference on Hardware/Software Codesign and System Synthesis (CODES+ISSS), 2014, pp. 10:1–10:10.

[11] M. Shafique, S. Garg, J. Henkel, D. Marculescu, The EDA challenges in the dark silicon era: temperature, reliability, and variability perspectives, in: 51st Design Automation Conference (DAC), 2014, pp. 185:1–185:6.

[12] M. Shafique, D. Gnad, S. Garg, J. Henkel, Variability-aware dark silicon management in on-chip many-core systems, in: The Conference on Design, Automation and Test in Europe (DATE), 2015.

[13] A. Kanduri, M.-H. Haghbayan, A.-M. Rahmani, P. Liljeberg, A. Jantsch, H. Tenhunen, Dark silicon aware runtime mapping for many-core systems: a patterning approach, in: 2015 33rd IEEE International Conference on Computer Design (ICCD), IEEE, 2015, pp. 573–580.

[14] N. Binkert, B. Beckmann, G. Black, et al., The gem5 simulator, SIGARCH Comput. Archit. News 39 (2) (2011) 1–7.

[15] C. Bienia, S. Kumar, J.P. Singh, K. Li, The PARSEC benchmark suite: characterization and architectural implications, in: Proceedings of the 17th International Conference on Parallel Architectures and Compilation Techniques (PACT), 2008, pp. 72–81.

[16] H. Khdr, S. Pagani, M. Shafique, J. Henkel, Thermal constrained resource management for mixed ILP-TLP workloads in dark silicon chips, in: Proceedings of the 52nd Annual Design Automation Conference, ACM, 2015, p. 179.

[17] J. Henkel, Y. Li, Energy-conscious HW/SW-partitioning of embedded systems: a case study on an MPEG-2 encoder, in: Proceedings of the 6th International Workshop on Hardware/Software Codesign, CODES/CASHE '98, IEEE Computer Society, Washington, DC, USA, 1998, pp. 23–27. ISBN 0-8186-8442-9, http://dl.acm.org/citation.cfm?id=278241.278292.

[18] S. Kobbe, L. Bauer, D. Lohmann, W. Schroder-Preikschat, J. Henkel, DistRM: distributed resource management for on-chip many-core systems, in: 2011 Proceedings of the 9th International Conference on Hardware/Software Codesign and System Synthesis (CODES+ISSS), 2011, pp. 119–128.

[19] I. Anagnostopoulos, V. Tsoutsouras, A. Bartzas, D. Soudris, Distributed run-time resource management for malleable applications on many-core platforms, in: 50th Design Automation Conference (DAC), 2013, pp. 168:1–168:6.

[20] G. Girão, T. Santini, F.R. Wagner, Exploring resource mapping policies for dynamic clustering on NoC-based MPSoCs, in: The Conference on Design, Automation and Test in Europe (DATE), 2013, pp. 681–684.

[21] M. Fattah, M. Daneshtalab, P. Liljeberg, J. Plosila, Smart hill climbing for agile dynamic mapping in many-core systems, in: 50th Design Automation Conference (DAC), 2013, pp. 39:1–39:6.

[22] E.L. de Souza Carvalho, N.L.V. Calazans, F.G. Moraes, Dynamic task mapping for MPSoCs., IEEE Des. Test Comput. 27 (5) (2010) 26–35.

[23] C.-L. Chou, U.Y. Ogras, R. Marculescu, Energy- and performance-aware incremental mapping for networks on chip with multiple voltage levels, IEEE Trans. Comput. Aided Des. Integr. Circuits Syst. (TCAD) 27 (10) (2008) 1866–1879.

[24] S. Wildermann, M. Glaß, J. Teich, Multi-objective distributed run-time resource management for many-cores, in: The Conference on Design, Automation and Test in Europe (DATE), 2014, pp. 221:1–221:6.

[25] J. Lee, N.S. Kim, Optimizing throughput of power- and thermal-constrained multicore processors using DVFS and per-core power-gating, in: 46th Design Automation Conference (DAC), 2009, pp. 47–50.

[26] A. Bonfietti, L. Benini, M. Lombardi, M. Milano, An efficient and complete approach for throughput-maximal SDF allocation and scheduling on multi-core platforms, in: The Conference on Design, Automation and Test in Europe (DATE), 2010, pp. 897–902.

[27] V. Hanumaiah, S. Vrudhula, K.S. Chatha, Performance optimal online DVFS and task migration techniques for thermally constrained multi-core processors, IEEE Trans. Comput. Aided Des. Integr. Circuits Syst. (TCAD) 30 (11) (2011) 1677–1690.

[28] A. Rezaei, D. Zhao, M. Daneshtalab, H. Zhou, Multi-objective task mapping approach for wireless NoC in dark silicon age. in: 2017 25th Euromicro International Conference on Parallel, Distributed and Network-based Processing (PDP), 2017, pp. 589–592, https://doi.org/10.1109/PDP.2017.12.

[29] J. Henkel, L. Bauer, J. Becker, O. Bringmann, U. Brinkschulte, S. Chakraborty, M. Engel, R. Ernst, H. Hrtig, L. Hedrich, A. Herkersdorf, R. Kapitza, R. Lohmann, P. Marwedel, W. Platzner, W. Rosenstiel, U. Schlichtmann, O. Spinczyk, M. Tahoori, J. Teich, N. When, H.J. Wunderlich, Design and architectures for dependable embedded systems. in: 2011 Proceedings of the Ninth IEEE/ACM/IFIP International Conference on Hardware/Software Codesign and System Synthesis (CODES+ISSS), 2011, pp. 69–78, https://doi.org/10.1145/2039370.2039384.

[30] J. Henkel, L. Bauer, H. Zhang, S. Rehman, M. Shafique, Multi-layer dependability: from microarchitecture to application level. in: 2014 51st ACM/EDAC/IEEE Design Automation Conference (DAC), ISSN 0738-100X, 2014, pp. 1–6, https://doi.org/10.1145/2593069.2596683.

[31] A.K. Coskun, T.S. Rosing, K.A. Whisnant, K.C. Gross, Temperature-aware MPSoC scheduling for reducing hot spots and gradients, in: The Asia and South Pacific Design Automation Conference (ASP-DAC), 2008, pp. 49–54.

[32] P. Kumar, L. Thiele, Thermally optimal stop-go scheduling of task graphs with real-time constraints, in: The 16th Asia and South Pacific Design Automation Conference (ASP-DAC), 2011, pp. 123–128.

[33] T. Ebi, M. Faruque, J. Henkel, TAPE: thermal-aware agent-based power econom multi/many-core architectures, in: IEEE/ACM International Conference on Computer-Aided Design - Digest of Technical Papers, 2009. ICCAD 2009, 2009, pp. 302–309.

[34] T. Ebi, D. Kramer, W. Karl, J. Henkel, Economic learning for thermal-aware power budgeting in many-core architectures, in: 2011 Proceedings of the 9th International Conference on Hardware/Software Codesign and System Synthesis (CODES+ISSS), 2011, pp. 189–196.

[35] H. Khdr, T. Ebi, M. Shafique, H. Amrouch, J. Henkel, mDTM: multi-objective dynamic thermal management for on-chip systems, in: Proceedings of the Conference on Design, Automation & Test in Europe (DATE), 2014, pp. 330:1–330:6.

[36] S. Buchwald, M. Mohr, A. Zwinkau, Malleable invasive applications, in: Software Engineering (Workshops)2015, pp. 123–126.

[37] W. Huang, S. Ghosh, S. Velusamy, K. Sankaranarayanan, K. Skadron, M.R. Stan, HotSpot: a compact thermal modeling methodology for early-stage VLSI design, IEEE Trans. Very Large Scale Integr. (VLSI) Syst. 14 (5) (2006) 501–513.

[38] Intel Corporation, Single-Chip Cloud Computer (SCC), 2009, www.intel.com/content/www/us/en/research/intel-labs-single-chip-cloud-overview-paper.html.

[39] R.E. Kessler, The alpha 21264 microprocessor, IEEE Micro 19 (2) (1999) 24–36.

[40] Intel Xeon Phi Coprocessor Datasheet, https://ark.intel.com/products/94709/Intel-Xeon-Phi-Processor-7210F-16GB-1_30-GHz-64-core December, 2017.

[41] A. Grenat, S. Pant, R. Rachala, S. Naffziger, 5.6 Adaptive clocking system for improved power efficiency in a 28nm x86-64 microprocessor, in: ISSCC2014, pp. 106–107.

[42] W. Huang, M.R. Stant, K. Sankaranarayanan, R.J. Ribando, K. Skadron, Many-core design from a thermal perspective, in: 45th Design Automation Conference (DAC), 2008, pp. 746–749.

[43] Intel Corporation, Intel Turbo Boost Technology in Intel CoreTM Microarchitecture (Nehalem) Based Processors, 2008. White paper.

[44] H. Khdr, S. Pagani, E. Sousa, V. Lari, A. Pathania, F. Hannig, M. Shafique, J. Teich, J. Henkel, Power density-aware resource management for heterogeneous tiled multicores, IEEE Trans. Comput. 66 (3) (2017) 488–501.

ABOUT THE AUTHORS

Heba Khdr is a Ph.D. student at the Chair for Embedded Systems (CES) in Karlsruhe Institute of Technology (KIT) in Germany. She received her B.Sc. in Informatics Engineering from University of Aleppo, Syria in 2005 with an excellent grade and the first rank. From 2005 until 2007 she worked as a software engineer in software company in Syria. From 2008 until 2010 she worked as an assistant in Aleppo University. In 2011 she did an equivalent master thesis at KIT and started her Ph.D. at the Chair for Embedded System (CES). Her research interests are thermal management and resource management in many core systems. In 2012 she received Research Student Award from KIT. She received Best Paper Award from IEEE/ACM International Conference on Hardware/Software Codesign and System Synthesis (CODES+ISSS) in 2014.

Santiago Pagani is currently a Staff Firmware Engineer and Team Lead at ARM Ltd., Cambridge, UK, where he runs an Agile firmware development team working on key components for the next generation of Mali GPU products.

He received his Diploma in Electronics Engineering from the Department of Electronics, National Technological University (UTN-FRBA), Argentina in 2010.

He received his Ph.D. in Computer Science from the Karlsruhe Institute of Technology (KIT) with "Summa cum Laude" in 2016. From 2003 until 2012, he worked as a hardware and software developer in the industry sector for several companies in Argentina, including 2 years as technical group leader. From 2012 until 2017, he worked as a research scientist (doctoral researcher and later post-doc) as part of the research staff at KIT. He received two Best Paper Awards (IEEE RTCSA in 2013 and IEEE/ACM CODES+ISSS in 2014), one Feature Paper of the Month (IEEE Transactions on Computers in 2017), and three HiPEAC Paper Awards. He received the 2017 ACM SIGBED "Paul Caspi Memorial Dissertation Award" in recognition of an outstanding Ph.D. dissertation.

His interests include embedded systems, real-time systems, energy-efficient scheduling, power-aware designs, and temperature-aware scheduling.

Muhammad Shafique (M'11, SM'16) is a full professor at the Institute of Computer Engineering, Vienna University of Technology (TU Wien), Austria. He is directing the Group on Computer Architecture and Robust, Energy-Efficient Technologies (CARE-Tech). He received his Ph.D. in Computer Science from Karlsruhe Institute of Technology (KIT) in January 2011. Before, he was with Streaming Networks Pvt. Ltd. where he was involved in research and development of advanced video coding systems for several years.

His research interests are in computer architecture, power- and energy-efficient systems, robust computing, hardware security, brain-inspired computing, neurosciences, emerging technologies and nanosystems, cognitive systems, FPGAs, and embedded systems. His research has a special focus on cross-layer analysis, modeling, design, and optimization of computing and memory systems, as well as their integration in IoT and CPS domains. Dr. Shafique received the 2015 *ACM/SIGDA Outstanding New Faculty Award*, six gold medals in educational career, and several best paper awards and nominations at prestigious conferences like DATE, DAC, ICCAD and CODES+ISSS, Best Master Thesis Award, and Best Lecturer Award. He has given several Invited Talks, Tutorials, and Keynotes. He is a senior member of the IEEE and IEEE Signal Processing Society (SPS), and a member of ACM, SIGARCH, SIGDA, SIGBED, and HiPEAC. He holds one US patent and over 180 papers in premier journals and conferences.

Jörg Henkel (M'95-SM'01-F'15) is currently with the Karlsruhe Institute of Technology (KIT), Germany, where he is directing the Chair for

Embedded Systems (CES). Prof. Henkel received the masters and the Ph.D. (Summa cum laude) degrees, both from the Technical University of Braunschweig, Germany.

He then joined the NEC Laboratories, Princeton, NJ, USA. His current research interests include design and architectures for embedded systems with focus on low power and reliability. Prof. Henkel has received the 2008 DATE Best Paper Award, the 2009 IEEE/ACM William J. Mc Calla ICCAD Best Paper Award, the CODES+ISSS 2011, 2014 and 2015 Best Paper Awards. He was the General Chair of major CAD events including ICCAD and ESWeek. He is the Chairman of the IEEE Computer Society, Germany Section, and was the Editor-in-Chief of the ACM Transactions on Embedded Computing Systems for two terms. He is currently the Editor-in-Chief of the IEEE Design\&Test Magazine. He is also an Initiator and Spokesperson of the national priority program on Dependable Embedded Systems of the German Science Foundation and the site coordinator (Karlsruhe site) of the three-university collaborative research center on invasive computing. He is a Fellow of the IEEE and holds 10 US patents.

CHAPTER FIVE

Dynamic Power Management for Dark Silicon Multicore Processors ☆

Siddharth Garg
New York University, Brooklyn, NY, United States

Contents

Abstract

Multicore processors in the dark silicon era will be over-provisioned with multiple cores of different types and will have to satisfy stringent power constraints while maximizing performance. As such, a critical research problem is that of developing new, effective, and scalable run-time dynamic power management techniques and algorithms to effectively make use of the over-provisioned but power-constrained computational resources on a chip. This chapter surveys the state of the art on this subject and dives in-depth into three specific dynamic power management solutions for multicore processors in the dark silicon era.

☆ Parts of this chapter describe the author's prior work in publications [1–3]. Consequently, some of the text and the figures in this chapter are adapted from these prior publications.

Advances in Computers, Volume 110
ISSN 0065-2458
https://doi.org/10.1016/bs.adcom.2018.03.016
© 2018 Elsevier Inc.
All rights reserved.

1. INTRODUCTION

Technology scaling enables a greater number of transistors, hence processing cores, to be integrated on a single chip by approximately doubling the number of transistors every 1.5 years. However, designs are now primarily limited by power and not area due to increasing power density, which results in the so-called dark silicon problem [4, 5].

The abundance of transistors in the dark silicon era enables multicore processors to be overprovisioned. That is, the chip is provisioned with redundant cores, more than the number that can be simultaneously operated at peak performance. Over-provisioned multicores can be either homogeneous (all cores are identical) or asymmetric/heterogeneous that integrate high performance "big" and more energy efficient "little" cores on the same chip [6–8]. In the latter case, the underlying idea is to use the small cores for lightweight tasks, and the big cores for computationally demanding tasks.

For both homogeneous and heterogeneous/asymmetric multicores, the problem of designing run-time power management strategies to maximize performance under power and thermal constraints is a critical one. Power management for homogeneous multicores builds upon power management strategies first proposed for single core processors, principally dynamic voltage and frequency scaling (DVFS) [9–12] and microarchitectural adaptation [12], by applying these to each core on the chip. For over-provisioned homogeneous multicores, the run-time system must also determine how many cores to turn on and a thread-to-core mapping, taking into account factors such as core-to-core variations in process, temperature, reliability, and aging. Run-time power management for heterogeneous multicores, in addition to the challenges described above, must take into account the relative power/performance of different core types and accordingly determine how many cores of each type to activate, and migrate threads from one core to another when required.

This chapter seeks to provide a comprehensive overview of dynamic power management strategies for over-provisioned multicore processors using three specific techniques as exemplars, and describing the state of the art for each. These are

- Microarchitectural adaptation for homogeneous multicores, discussed in Section 2.
- Dynamic scheduling for asymmetric multicores, discussed in Section 3.
- Dynamic degree of parallelism (DoP) and cluster migration for asymmetric multicores, discussed in Section 4.

2. DARK SILICON AWARE MICROARCHITECTURAL ADAPTATION FOR HOMOGENEOUS MULTICORES

Microarchitectural adaptation enables the microarchitectural configuration of each core to be adjusted at run-time. This includes microarchitectural parameters such as issue width, reorder buffer size, cache capacity, and others. Recent results have shown that microarchitectural adaptation generalizes coarse-grained dynamic core-count scaling [13], and as a consequence is more effective. Fig. 1 shows a general purpose processor with 4-cores in which the instruction fetch queue, number of execution units and L1-D/L1-I cache sizes can be dynamically changed using power gating. Note, in the figure, that a portion of each core is dark. Summed over all cores, the run-time microarchitectural solution must ensure that enough of the chip is kept dark so as to meet power constraints.

To perform fine-grained dynamic adaptation, a challenging global optimization problem must be solved by the run-time power manager, i.e., how to determine the configuration of each core so as to maximize performance within the power budget. The problem is challenging for three reasons: (i) the solution must be scalable, particularly for multicore systems that have tens or even hundreds of cores; (ii) the complex relationship between different core configurations and the corresponding power/performance of

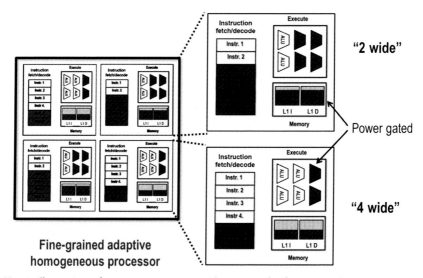

Fig. 1 Illustration of a 4-core processor with support for fine-grained synamic microarchitectural adaptation.

the thread running on the core must be modeled and predicted, and (iii) for multithreaded applications, the run-time power manager must also determine how speeding up a single thread will affect the performance of the application as a whole.

2.1 State of the Art

The two primary control knobs for dynamically adaptive processors are DVFS and microarchitectural adaptation. Herbert and Marculescu [14] empirically studied the design space of fine-grained DVFS for multicore processors and evaluated different V/F domain granularities and a number of power management algorithms, with sum-IPS per Watt as the performance metric. Isci et al. [15] performed a similar study, but with the goal of maximizing performance within a power budget, and use MaxBIPS as their measure of performance. Others have studied the scalability of the global optimization problem and proposed distributed approaches for reduced complexity [16, 17]. A related line of research has studied the use of control theory for power-constrained performance maximization [18, 19]. Cochran et al. [20] present a machine learning based approach based on offline workload characterization (and online prediction) but perform DVFS adaptation at a coarse time granularity of 100 billion uops. Recently, Godycki et al. [21] have proposed reconfigurable power distribution networks to enable fast, fine-grained, per-core voltage scaling and use this to *reactively* slow-down stalled threads and redistribute power to working threads.

Rangan et al. [22] propose thread motion, fine-grained migration of threads from one core to another on an asymmetric multicore processor, in order to emulate DVFS without its overheads. Others have focused on leveraging asymmetry to increase the performance of multithreaded applications by identifying and accelerating critical sections [23–27]. A more recent work by Craeynest et al. [28] proposes to use fairness-aware equal-progress scheduling on heterogeneous multicores.

The state of the art on microarchitectural adaptation of multicore processors spans ideas ranging from core-level to fine-grained power gating [13, 29–32]. Core fusion [33] and core morphing [34] enable two or more cores to be fused at run-time to enhance single thread performance. In this chapter, we will discuss in detail thread progress equalization (TPEq), a recently proposed schemed for power-constrained fine-grained microarchitectural adaptation for multithreaded workloads [1].

2.2 Thread Progress Equalization

Instructions per second (IPS) is a direct indicator of performance for sequential, single-threaded applications. Moreover, for multiprogrammed workloads, the IPS summed over all threads indicates net throughput and is a commonly used performance metric [14, 15, 35]. However, for multithreaded applications, the sum of IPS metric can be misleading. For instance, a thread that is spinning on a lock or waiting at a barrier might execute usermode synchronization instructions, but these do *not* correspond to useful work. In fact, speeding some threads may not have an appreciable impact on application execution latency. The problem is heightened by the fact that programmers exploit parallelism in different ways—for example, using data-level parallelism with barrier synchronization, or task-level parallelism with local/global task queues and static/dynamic load balancing.

Thread progress equalization (TPEq) is a run-time mechanism to maximize performance within a power budget for multithreaded applications running on multicore processors with per-core dynamic adaptation. The basic idea behind TPEq is that all threads that are expected to synchronize on a barrier should arrive at the barrier at the same time so as to best utilize the available power budget. It is easily seen why this makes sense by using a contradiction-based argument. That is, if the statement above is not true, *early* threads can be slowed down and the power saved by doing so can be allocated to speed-up *lagging* threads, getting better latency.

Threads can be imbalanced for two reasons. First, even threads executing the same code but on different data inputs can have different instructions per cycle (IPC) counts. For example, the sequence of data accesses that one thread makes can have less spatial locality than another thread's accesses, resulting in more cache misses and lower IPC for the first thread. This is referred to as *IPC heterogeneity*. Second, threads might execute a different number of instructions until they reachs a barrier. This is because of imperfect load balancing or because the control flow of each thread might be different. This is referred to as *instruction count heterogeneity*.

Fig. 2 shows an example of two benchmark applications, FFT and Water. Nsq (SPLASH-2 [36]), executing on a homogeneous multicore processor. FFT exhibits only IPC heterogeneity. Each thread executes exactly the same number of instructions between barriers. Water.Nsq exhibits both IPC and instruction count heterogeneity, which can be seen from different slopes of threads in progress plot, *and* instruction count heterogeneity. Note that over the entire length of the application, thread T16 executes more than 1.15 × the number of instructions compared to thread T7.

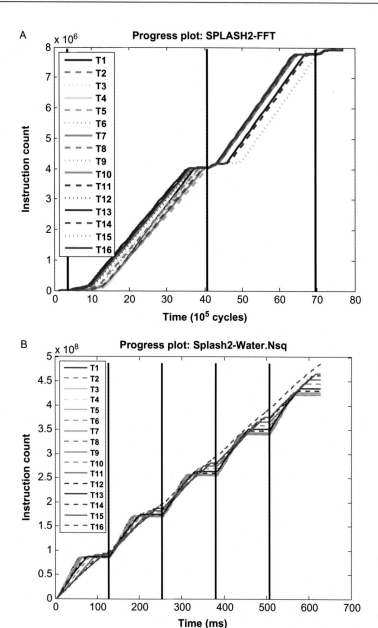

Fig. 2 Progress plots for the FFT and Water.Nsquared (SPLASH-2) benchmarks with 16 threads on a 16-core architecture. The *solid vertical lines* indicate barriers. *Slope* corresponds to IPS of that thread and hence the flat regions indicate time periods when the thread is stalled waiting for lagging threads to arrive. (A) IPC heterogeneity only. (B) IPC and instruction count heterogeneity.

TPEq seeks to dynamically optimize the configuration of each core/ thread such that each threads reach barriers at the same time by simultaneously accounting for *both* IPC and instruction count heterogeneity TPEq contains two basic components:

TPEq optimizer: given an oracle that can predict (i) the IPC and power consumption of each thread for every core configuration, and (ii) the total number of instructions the thread must execute until the next barrier, TPEq uses an efficient *polynomial-time* algorithm that *optimally* determines the core configuration for each thread to maximize application performance under power constraints.

TPEq Predictors: TPEq implements accurate run-time predictors for (a) IPC and power consumption of a thread for different core configurations, and (b) the number of instructions each thread executes between barriers.

2.2.1 TPEq Design and Implementation

Fig. 3 shows an overview of the design of TPEq. The hardware platform consists of a dynamically adaptive multicore architecture where, for example, each core can have a variable ROB size and the fetch width. In general, assume that each of the N cores can be set in one of M different *configurations* as described in Table 1. In its current implementation, TPEq assumes that the number of threads equals the number of cores, and a static mapping of threads to cores [37].

The TPEq run-time system consists of two components. The TPEq predictors monitor on-chip performance counters and predict the future application characteristics. The predictions are passed on to the TPEq optimizer, which determines the optimal configuration of each core so as to maximize overall system performance within a power budget. We now describe the design and implementation of TPEq.

2.2.1.1 TPEq Optimizer

The TPEq optimizer is at the heart of TPEq approach. To understand how the optimizer works, assume that we begin at the time instant when N threads exit a barrier and start making progress toward the next barrier. The optimal configuration for each core/thread needs to be decided for the interval between these two successive barriers. Assume that an oracle provides access to the following information:

- The number of instructions each thread executes until it enters the next barrier is in ratio $w(1) : w(2) : \ldots : w(N)$. Note that $w(1), w(2), \ldots, w(N)$

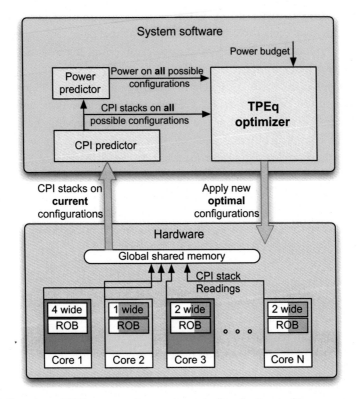

Fig. 3 Overview of TPEq approach on a dynamically adaptive multicore processor.

can be absolute instruction counts, but we only require the number of instructions each thread executes relative to other threads.

- The CPI of thread i ($1 \leq i \leq N$) when it executes on a core with configuration j ($1 \leq j \leq M$) is CPI(i, j), and the corresponding power dissipation is $P(i, j)$. We assume, for now, that for a given core configuration, the CPI and power dissipation of each thread do not change with time, at least until it reaches the next barrier. This assumption is relaxed later.

Under the assumptions above, TPEq tries to assign a configuration to each core/thread so as to stay within power budget P_{budget}, while minimizing the time taken by the most lagging thread to reach the next barrier. A key contribution of TPEq is an algorithm that *optimally* solves this problem in $\mathcal{O}(MN\log N)$ time.

The algorithm works as follows: TPEq starts by setting all cores to the configuration that consumes the least power and determines the identity of the *most lagging* thread for this setting, i.e., the thread that would reach

Table 1 Microarchitectural Adaptation Configurations

Configuration	Dispatch Width	ROB Size	Integer ALUs
1	1	16	1
2	2	32	3
3	2	64	3
4	4	64	6
5	4	128	6

Number of cores. 16, number of threads: 16 (1 thread/core)
Frequency: 3.5 GHz, voltage: 1.00 V, 22nm technology node
L1-I cache: 128 KB, write-back, 4-way, 4-cycle
L1-D cache: 128 KB, write-back, 8-way, 4-cycle
L2 cache: private 256 KB, write-back, 8-way, 8-cycle
L3 cache: 8 MB shared/4 cores, write-back, 16-way, 30-cycle
Cache coherence: directory-based MSI protocol
Floating point units: 2, Complex ALUs: 1

the barrier last. For thread i, the number of clock cycles required to reach the barrier when executing on configuration j would be $w(i)\mathrm{CPI}(i,j)$. We define the *progress* of this thread as:

$$progress(i) = \frac{1}{w(i)\mathrm{CPI}(i,j)}$$

to capture the intuition that larger values of "progress" are better.

The configuration of the most lagging thread is then moved up to the next level,[a] and the new most lagging thread is determined. The core configuration for this new most lagging thread is now moved up by one level, and so on. This continues until there is no core whose configuration can be increased to the next level without violating the power budget. The resulting core configurations are optimal in terms of total execution time and are then updated in hardware. Algorithm 1 is a formal description of this optimization procedure.

---

[a] The configurations are, by convention, sorted in ascending order of power consumption. Also, TPEq limits the search to *Pareto optimal* configurations, by simply discarding ones where increasing power does not lead to increased performance.

ALGORITHM 1 TPEq Optimization Procedure

1 $P_{tot} \leftarrow 0$;
  // Init. all threads to lowest core config.
2 **for** $i \in [1, N]$ **do**
3     $c(i) \leftarrow 1$;
4     $P_{tot} \leftarrow P_{tot} + P(i, c(i))$;
5     $progress(i) \leftarrow \frac{1}{w(i)\,CPI(i,c(i))}$;
6 **end**
7 **while** $P_{tot} \leq P_{budget}$ **do**
      // Determine lagging thread l
8     $l \leftarrow arg\,min_{i \in [1,N], c(i) < M} \left\{ \frac{1}{w(i)\,CPI(i,c(i))} \right\}$;
      // If no such thread exists
9   **if** $l = \emptyset$ **then**
10   break;
11   **end**
      // Increase core configuration of lagging thread
12     $c(l) \leftarrow c(l) + 1$;
      // Update progress and power
13     $progress(l) \leftarrow \frac{1}{w(l)\,CPI(l,c(l))}$;
14     $P_{tot} \leftarrow P_{tot} - P(l, c(l) - 1) + P(l, c(l))$;
15 **end**
  // Return optimal core configurations
16 **return** c;

In practice, the TPEq optimization routine is called once every *epoch* in order to address fast variations in thread characteristics. The epoch length ($\mathcal{E}$, measured in number of clock cycles) is configurable. The epoch length should be short enough to quickly adapt to CPI and power variations, but is practically limited by the computational overhead of the optimization procedure.

2.2.1.2 TPEq Predictors
In Section 2.2.1.1 it was assumed that the TPEq optimizer has oracular knowledge of the relative instruction counts of the threads. In practice, the TPEq predictors determine these values at run-time for each thread

immediately after a synchronization related stall. TPEq also requires predictions for CPI and power consumption of each thread for every core configuration once every *epoch*, i.e., in synchrony with the TPEq optimization procedure.

TPEq's instruction count predictor predicts the number of instructions each thread executes relative to other threads. The predictor is based on the observation that *the number of instructions each thread executes between barriers, relative to other threads, remains the same*. This motivates the use of a history-based predictor to predict relative instruction counts.

Intuitively, the difference in relative instruction counts of several multi-threaded workloads arise as a result of imbalance in the amount of computing for a thread, which persists across several barriers. Singh et al. [38] were perhaps the first to qualitatively observe the locality in the data distribution in threads across successive barriers in many of our benchmark algorithms and provide insights into this characteristic. They noted that successive barriers correspond to very small "time-steps" in the physical world, and that the characteristics of the physical world change slowly with time. Hence, the amount of work to be performed by a thread in one time-step, is a good predictor for the amount of work in the next time-step. In the progress plot for Water.Nsq (see Fig. 2B), for instance, the number of water molecules per thread remain nearly constant across barriers. Consequently, thread T16 (T7), with most (least) number of water molecules, always executes the most (least) instructions in any interbarrier interval.

Quantitatively, this trend can be observed over several barrier synchronization-based benchmarks in the SPLASH-2, PARSEC, and Phoenix benchmarks suites. Fig. 4 shows the a scatter plot of relative instruction counts in barrier phase $t + 1$ vs the relative instruction counts of threads in barrier phase t across all benchmarks with instruction count heterogeneity (coded in different colors). The mean absolute relative error using a last-value-predictor for relative instruction counts was found to be only 4.2%. Liu et al. [39] have observed similar locality behavior across the outermost loops of the SpecOMP parallel applications, and use last-value prediction to perform voltage/frequency scaling for each thread.

TPEq's CPI and power predictors operate as follows. Let $CPI_t(i, j)$ be the CPI of thread i on core configuration j in epoch t. The goal of the CPI predictor is to determine $CPI_{t+1}(i, j)$ for all $j \in [1, M]$. Duesterwald et al. [40] have shown that for predicting the CPI in the next epoch assuming the same core configuration, i.e., characterizing temporal variability in CPI, last-value

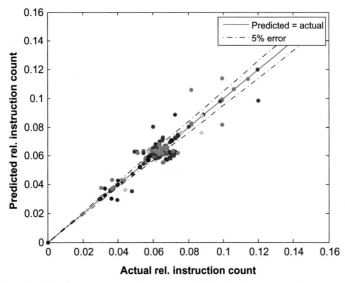

Fig. 4 Scatter plot of predicted and actual relative instruction counts between successive barriers for different benchmarks.

predictors perform on a par with exponentially weighted mean, table-based, and cross-metric predictors. The accuracy of last-value predictors improves for shorter prediction epochs. The last-value predictor simply implements:

$$\mathrm{CPI}_{t+1}(i,j) = \mathrm{CPI}_t(i,j).$$

To predict $\mathrm{CPI}_{t+1}(i, k)$ for all $k \neq j$ given $\mathrm{CPI}_{t+1}(i, j)$, TPEq needs an approach that predicts the performance on one core type given performance on another core type. For this, TPEq uses CPI stack information measured using hardware counters broken down into four components: compute CPI (base CPI in the absence of miss events), memory CPI (cycles lost due to misses in the memory hierarchy), branch CPI (cycles lost due to branch misprediction), and synchronization CPI (cycles lost due to stalls on synchronization instructions).

With these measurements on configuration j, TPEq predict the CPI on configuration k using a linear predictor as follows:

$$\mathrm{CPI}_t(i, k) = \alpha_{jk}^{\mathrm{comp}} \mathrm{CPI}_t^{\mathrm{comp}}(i, j) + \alpha_{jk}^{\mathrm{mem}} \mathrm{CPI}_t^{\mathrm{mem}}(i, j)$$
$$+ \alpha_{jk}^{\mathrm{branch}} \mathrm{CPI}_t^{\mathrm{branch}}(i, j) + \alpha_{jk}^{\mathrm{synch}} \mathrm{CPI}_t^{\mathrm{synch}}(i, j).$$

The pairwise α_{jk}^* parameters, one for every pair of core configurations, are learned offline using training data obtained from a set of representative benchmarks and stored for online use. Note that the learned parameters are not benchmark specific and depend only on the core configurations.

A similar linear predictor that utilizes CPI components was proposed by Lukefahr et al. [41], although only for big-little core configurations. Another CPI predictor is PIE [42], which makes use of information collected using hardware counters including the total CPI, CPI of memory instructions, misses per instruction (MPI), and data dependencies between instructions. However, PIE has been proposed for CPI prediction between small in-order and large out-of-order cores, while TPEq also requires predictions between different out-of-order core configurations and also faces the challenge of predicting over future epoch. Furthermore, since training the TPEq predictor is automated and data-driven, it is easy to deploy for a large number of core configurations.

We note that existing processors such as the Intel Pentium 4 [43] and the IBM POWER5 [44] have built-in hardware support for performance counters that measure CPI components. In addition, Eyerman et al. [45] have proposed a performance counter architecture that further improves upon the accuracy of these commercial implementations with similar hardware complexity. Their approach provides very accurate estimates of the CPI stack components with only 2% average absolute error.

The TPEq power predictor uses the predicted CPI values for each core configuration (as described in Fig 5) to predict their power consumption. This is based on previous work which indicates that CPI (or IPC) is highly correlated with power consumption [46, 47]; for instance, Bircher and John report on average only 3% error when compared to measured CPU power [47]. Thus, the TPEq power predictor predicts the power consumption for different core types as follows:

$$P(i,j) = \beta_{0,j} + \frac{\beta_{1,j}}{\mathrm{CPI}(i,j)} + \frac{\beta_{2,j}}{\mathrm{CPI}(i,j)^2}$$

where $\beta_{0,j}$, $\beta_{1,j}$, and $\beta_{2,j}$ are fixed parameters that are learned for each core type offline and stored for online use.

2.2.2 Empirical Analysis of TPEq

TPEq is empirically evaluated on the Sniper [48] multicore simulator for x86 processors. The baseline is processor with 16 cores and an 80 W power budget. The relevant core/uncore microarchitectural parameters are shown in Table 1. Each core can pick from one of five different configurations which are also listed in Table 1. The issue queue and load-store queue are scaled

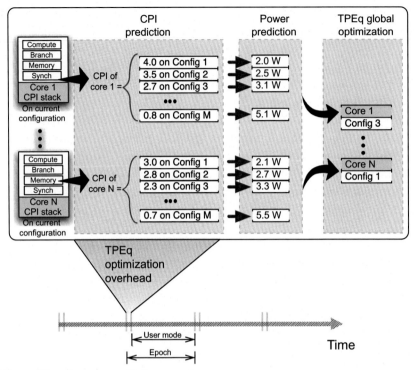

Fig. 5 CPI and power prediction overview.

Table 2 Maximum IPC and Power Observed for Different Configurations Using Swaptions

	Conf. 1	Conf. 2	Conf. 3	Conf. 4	Conf. 5
IPC	0.65	1.06	1.13	1.27	1.36
Power (W)	3.93	5.43	5.56	6.51	6.69

automatically with ROB size, since in Sniper all three are governed by a single parameter "*window size*"). Table 2 shows the maximum observed IPC and power values over all epochs for different static core configurations using Swaptions benchmark. Finally, in all experiments, the epoch length is set to 1 ms (3.5 million clock cycles at the baseline clock frequency of 3.5 GHz).

Fig. 6 shows TPEq progress plots for the FFT benchmark along with those for MaxBIPS [15], a technique that maximizes the sum of IPS over all threads. MaxBIPS is a commonly used (and intuitive) objective for applications where the threads are independent—multiprogrammed workloads,

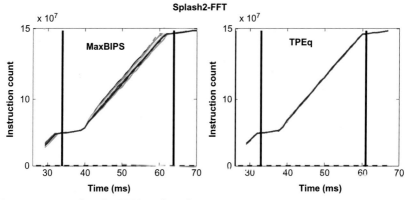

Fig. 6 Progress plots for FFT benchmark.

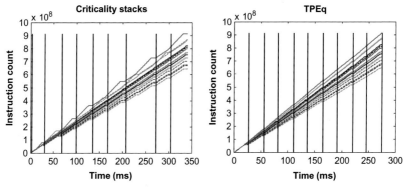

Fig. 7 Progress plots for fluidanimate using CS and TPEq.

for example. Compared to the baseline FFT progress plot in Fig. 2A, both the MaxBIPS and TPEq progress plots have lower heterogeneity in thread progress and stall far less frequently. Between the two, TPEq is more effective than MaxBIPS. TPEq achieves an almost 60% *reduction in stall time* compared to the baseline without microarchitectural adaptation. Overall, these results point to the importance of optimizing for the right objective function, as TPEq does, during run-time adaptation.

Next TPEq is compared with another approach explicitly designed for reducing stall time in multithreaded workloads—criticality stacks (CS) [24]. The benchmark used is fluidanimate, whose progress plots are shown in Fig. 7. TPEq is more successful in reducing thread stalls (regions where a thread's progress plot is flat) than CS, primarily because TPEq speeds-up or slows-down each thread optimally so they reach barriers at (about) the same time, while CS only speeds-up the most critical thread. In fact, the most critical thread identified by CS is speed-up more than necessary, and end up stalling on the next barrier. Compared to the baseline, in which all threads are executed on identical cores within same power budget, TPEq reduces total stall time by as much as 50%, while CS only results in less than 20% reduction in stall time.

Further insight can be obtained from Fig. 8, which shows the time spent by each thread in each configuration for CS and TPEq. Although it is clear that both CS and TPEq identify Thread 11 as most critical (assigning it to higher power/performance configurations), TPEq assigns each thread (including Thread 11) to a great range of configurations since it is able to perform fine-grained optimization. In fact, configuration 4 is not utilized by CS at all, while this is not the case for TPEq.

From a dark silicon stand-point that each thread spends most time in Config. 2, the second lowest power configuration. In this configuration, a core's dispatch width is reduced by 50%, 50% of the ALUs are dark, the ROB size is a fourth of the maximum ROB size. Of course, not all cores are necessarily in Config. 2—some cores utilize the power slack introduced by other cores running in low power configuration by running in higher power modes.

3. DYNAMIC SCHEDULING FOR ASYMMETRIC MULTICORES

Processor designers are increasingly adopting the asymmetric multi-core paradigm, in large part because of the challenges posed by dark silicon. Fig. 9 is an example of a commercial octa-core ARM processor, which has four "big" cores and four "little" cores. However, there is a critical need to devise new dynamic scheduling policies for such platforms that can most effectively utilize the available cores without exceeding power or thermal budgets.

Indeed, conventional scheduling policies that focus only on performance, for instance, the default scheduling policy on the commercial big-LITTLE platform that opportunistically uses more big cores as jobs arrive, result in system shutdown since the thermal budget is quickly surpassed. (This happens *despite the fan*, as depicted in Fig. 10.) An somewhat better solution is to utilize the power efficient little cores first, and start using the big cores only after all the little cores are fully utilized. Although this

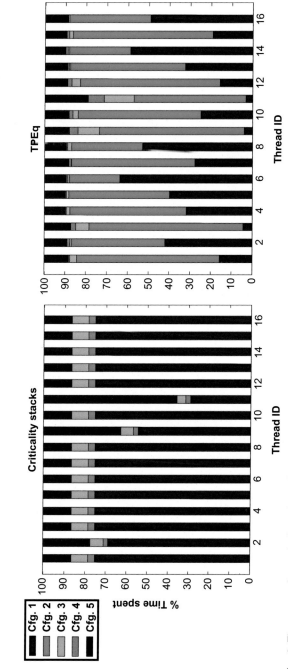

Fig. 8 Time spent by each thread in different configurations for CS and TPEq.

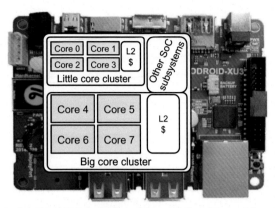

Fig. 9 The state-of-art big-LITTLE platform [49], with four big and four little cores.

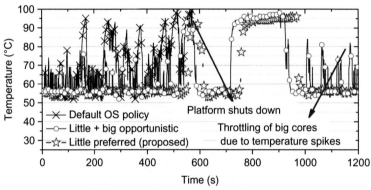

Fig. 10 The baseline policies (default OS policy and little + big opportunistic) result in either system shutdown or performance throttling due to over-utilization of the big cores. In contrast, the proposed policies result in thermally safe operation with significant benefits in both performance and energy efficiency.

prevents system shutdown, transient temperature spikes, shown in Fig. 10 (blue o markers), cause the system governor to throttle big cores, thus reducing performance. In contrast, the dark silicon aware policy proposed by Jain et al. [2], and described herein, avoid both system shut down and performance throttling by *judiciously* determining when to use big cores, as seen in Fig. 10 (red ⋆ markers).

The dark silicon aware scheduling policy is described in the context of an asymmetric processor like the ARM big.LITTLE deployed in a datacenter to a stream of requests, for instance search queries. The goal is to meet prespecified service-level agreements (SLA), for instance, on average latency with which a request is served. This is also referred to as the mean sojourn time. The rate at which queries arrive, the arrival rate, can vary significantly with time which, as we will see, affects the utilization of big cores.

When the goal is to minimize mean sojourn time (the sum of the time the job spends waiting in the queue and executing on a core), it would seem intuitive to schedule jobs to *any* core, little or big, that is available to exploit all available computing resources. However, seminal work from Lin and Kumar [50] has shown that the optimal scheduling policy is of *threshold type*. That is, little cores are used only if the number of outstanding requests in the queue is above a certain threshold. This is referred to as a threshold policy.

With added power and thermal constraints, little cores become preferred because they are more power efficient. A threshold policy in this context will always assign jobs to a little core when one is available, and use big cores only if the number of outstanding jobs in the scheduler's queue exceeds a threshold. In this chapter, we describe the work of Jain et al. [2] that explores the design space of such threshold policies for ARM big.LITTLE processors.

3.1 Related Work

Several studies have looked at scheduling policies for homogeneous multi-cores [51–53] assuming tasks with a uniform arrival rate. For instance, the min-min policy prioritizes jobs with minimum execution time [54] for scheduling. Several algorithms additionally take energy consumption into account, for instance [55].

Since then several research efforts have been dedicated to determining run-time scheduling techniques to map application threads to cores [56–58]. However, none of these techniques consider dynamic job arrivals or queueing effects, which are critical in the server/datacenter context. Closer to our work, Gupta and Nathuji [59] have looked at datacenter servers with many little cores and one big core, and have modeled service time using M/M/1 model, i.e., assuming only one job runs on the server at any given point in time. In contrast, our work allows concurrent execution of multiple jobs, for instance, search queries with multiple little and big cores. The work in [60] looks at a similar problem, but assumes only one cluster is on at a given time. However, the entire processor is modeled as a single server running only one job at any given time, where the jobs itself are multithreaded and the big core is used to accelerate serial sections of the jobs.

With the introduction of commercially available advent of asymmetric or heterogeneous multicores like the ARM big.LITTLE architecture, there has been increased interest in designing performance optimal and energy efficient scheduling policies for heterogeneous processors. Recent work has proposed energy efficient scheduling policies for mobile web browsing on an ARM big.LITTLE processor [61, 62]. These techniques schedule jobs to exploit the variations in energy consumption and time required to load different web pages. Power/performance modeling of asymmetric multi-core architectures has been addressed in [63]. The authors present a software-based approach to estimate power consumption and performance for different core types. Likewise, power management of big-little processors is considered in [7], but is focused on dynamic core-count and voltage/frequency scaling during the execution of a single job, but does not model job arrivals or queuing effects. However, this work focuses on controlling the set of active cores, voltage/frequency of the cores and task migration. Similar to the previous papers on scheduling for heterogeneous processors, it does not consider a server setting and hence job arrival process and queuing effects are not modeled.

From a theoretical perspective, Walrand [64] and Koole [65] provide alternative (and simpler) proofs for the results of Lin and Kumar. The threshold-based policies are proven to be energy efficient with N thresholds policy for multiple servers in [66] with comparison of different policies, where it is shown theoretically energy efficiency gain is achieved with 1 or 2 thresholds. Nonetheless, extending the result showing the optimality of the threshold policies to settings with more than two servers has proven surprisingly hard.

Efficiency of big and small cores in web search has been analyzed in [67]. Finally, we note that heterogeneity in processing capabilities has also been discussed at the full datacenter level [68, 69], instead of a single server/processor level as we do in this chapter. These works do model job arrivals and queuing effects, but consider entirely different types of scheduling policies than those that this thesis research consider.

In this dark silicon era, we should also consider temperature aware scheduling algorithms with energy optimizations as thermal constraints play a important role. There are algorithms which do use task migration to get uniform thermal map based on HotSpot thermal model [70] or by using dynamic voltage/frequency scaling (DVFS) and dynamic thermal management (DTM) or heuristic algorithms [71]. Dynamic thermal management primarily by throttle the processor activity by either reducing frequency (or voltage) or restricting the operation of a core.

3.2 Threshold Policies for Asymmetric Multicores

We start off with the scenario in which the scheduler knows only the mean execution times on the big and little cores, and service jobs in the order in which they arrive (these constraints will be relaxed later). Under this assumption, when there is at least one idle core and one job waiting in the queue, the scheduler can make one of the following two decisions:

- *Send:* The job at the head of the queue is sent to an available core. If both types of cores are available, the scheduling policy decides to send the job to the *preferred* cluster. In general, the preferred cluster can be either the big cluster or the little cluster.
- *Wait:* The scheduling policy chooses to wait rather than scheduling a pending job to an available core.

Intuitively, it might seem that performance is improved by exploiting any available server, i.e., by always taking the *"send"* action. When the goal is to minimize the service time, the scheduler can utilize first all the big cores, and then use the little cores—this is, *in essence, what the default OS scheduler*

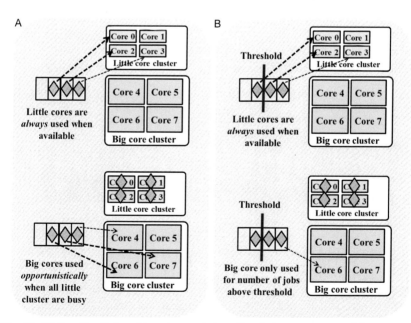

Fig. 11 (A) Baseline little + big opportunistic policy: little cores are used first and big cores when the all little cores are busy. (B) Threshold policy: little cores are used first, and big cores only for the number of jobs beyond a threshold.

does. However, policies in which big cores are preferred are *not* thermally stable, as shown in Fig. 10.

Therefore, the *baseline policy* first utilizes the little cores, and starts using the big cores only after all little cores are busy, as shown in Fig. 11-A. Meanwhile, any unused core is put to sleep state to save power. Big cores are used only when the number of jobs in the queue exceeds a threshold. The proposed threshold policy is illustrated in Fig. 11. If there is a job waiting at the head of the queue *and* a little core is available (*Case A*), the job is *always* sent to the little core. If all little cores are busy *and* there is at least one more job waiting in the queue, the job is sent to a big core *only if* the queue length is greater than a threshold *t*. Otherwise, the job is held in the queue.

The threshold needs to be set to the value that minimizes the mean service time for a given arrival rate. However, the optimal threshold varies with arrival rate. Thus, the threshold is updated to track the dynamic variations in arrival rate. This is achieved by first characterizing the optimal threshold as a function of a constant rate, and storing this information. Since the optimal threshold does not change abruptly, the optimal thresholds are obtained for five different arrival rates that cover the whole range in the empirical data as shown in Fig. 12.

For many workloads including web search, the execution time of a job can be predicted ahead of time based on its characteristics. For instance, a

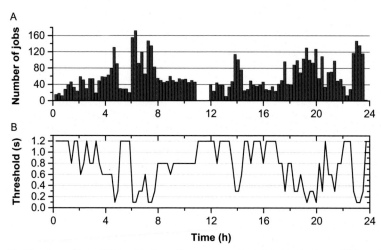

Fig. 12 Number of jobs (histogram) per hour production datacenter trace [73], and the optimal threshold policy (time-based threshold) with varying arrival rate.

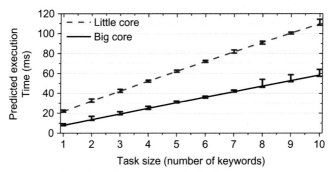

Fig. 13 Measured and predicted execution time for Clucene [72].

search query execution time predictor, which makes predictions based on the number of keywords in the search query, is presented in [74]. Fig. 13 shows the accuracy of keyword-based search query time predictors for the Clucene search engine [72]. Execution time prediction enables two additional optimizations, as detailed follows.

Time (Instead of Task)-Based Threshold: The threshold, so far, was expressed in terms of the number of tasks waiting in the queue. Using execution time prediction, the threshold can be expressed in terms of absolute time instead of number of tasks. In other words, the total execution time of jobs is computed and compared with the time-based threshold to determine when to schedule jobs on the big cluster. The time-based threshold is set, as before, to minimize mean response time for a given arrival rate and can be dynamically varied to respond to varying arrival rates.

Deadline-Aware Out-of-Order Execution: Assuming that the scheduler is allowed to execute jobs out-of-order, execution time prediction can be used to identify the most critical task in the queue. This is the task with the smallest difference between the deadline on the one hand, and the sum of its current waiting time and predicted execution time on the other. Thus, when scheduling a task to the big core, the scheduler preferentially executes the most critical job instead of the job at the head of the queue.

In conclusion, the family of scheduling policies explored can be configured depending on the following choices:
- Big or little preferred. Because big preferred policies are thermally unstable, only little preferred policies are explored.
- Static or dynamic threshold, i.e., if a static threshold is used or dynamically updated.
- Task or time-based threshold, i.e., whether the threshold is in terms of number of tasks or absolute execution time.

- In-order or out-of-order execution, i.e., whether the next job is scheduled from the head of the queue, or based on criticality to meet SLA.

We note that the in-order version of the algorithms are explicitly targeted toward minimizing mean response time, although empirically they also result in a significant increase in the fraction of queries serviced within the deadline. The out-of-order versions additionally aim to optimize for the SLA, but can only be used if the execution time can be predicted in advance. As a final note, the proposed scheduling policies are both implicitly and explicitly thermally constrained. Implicitly, the power governor on the chip throttles the big core or shuts off the platform in case of thermal emergencies. Explicitly, in line 10 of the algorithm, we only schedule jobs on the big core if the current system temperature is below a safe value.

3.2.1 Empirical Evaluation

The benchmark used to evaluate the proposed policy is the Clucene search engine [72], a C++ port of the Java-based open-source, high-performance text search engine API used in several websites such as Twitter, Wolfram Research, and LinkedIn [75]. Upon receiving a new job, the servers search as many keywords as specified in the job size in a 500 MB database from Wikipedia, already indexed using Clucene API. For each experiment, jobs are generated using entire day's job arrival data, taken from a commercially deployed datacenter server [73], and run on the system until all keyword searches are completed.

The Clucene search engine is executed on a Odroid XU3 development board [49] running Linux kernel 3.10.9 on Samsung Exynos 5422 MpSoC. Samsung Exynos 5422 MpSoC has a heterogeneous multiprocessing (HMP) solution with Big (4 A15 cores) and Little (4 A7 cores) clusters equipped with internal power and temperature sensors. The processor is widely used commercially in Android mobile phones and tablets.

The proposed scheduling policies are evaluated with the slow server (little core cluster) set as the preferred core type. These are compared against two baseline cases: (i) *only little*: The scheduling threshold is set to infinity such that only the little cluster is used, and (ii) *little + big opportunistic*: The threshold is set to zero such the big cores are used opportunistically whenever the little cores are fully utilized (Fig. 11A).

The threshold parameter, either in terms of number or predicted execution time of outstanding jobs, is an important knob that determines the utilization of the big cores for the proposed policies. Fig. 14 shows the performance and energy efficiency of the time-based threshold policy as a

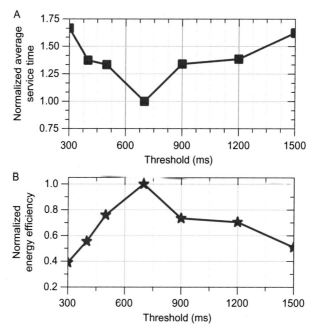

Fig. 14 Normalized average service time and energy efficiency for static out-of-order scheduling as a function of threshold expressed in absolute time with the optimal points as 10 s and 94 tasks/J, respectively. Both plots demonstrate the optimal nature of the "threshold"-based scheduling algorithms. Normalized average service time (A) and energy efficiency (B).

function of the threshold. Increasing the threshold results in lower utilization of the big cluster and should, in theory, cause the average service time to *increase* monotonically. However, note that increasing the scheduling threshold from 300 to 700 ms *improves* the average service time. This is because high utilization of the big cluster at lower thresholds results in thermal spikes that cause the big cluster frequency to be throttled, hence degrading performance. Further increasing the threshold beyond 700 ms hurts performance because of decreasing big cluster utilization. At the same time, the average power consumption decreases monotonically with threshold. Therefore, the energy efficiency peaks at the same threshold as performance, increasing from only 36 tasks/J at a threshold of 300 ms to 93 tasks/J at the optimal threshold.

Next, the proposed threshold policies are compared to the two baseline policies in terms of performance, power, and energy efficiency. Average service time for the baseline and proposed policies is plotted in Fig. 15A. Using

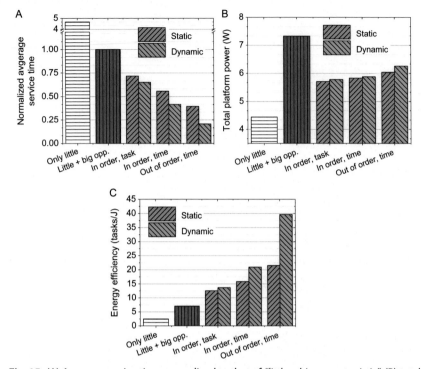

Fig. 15 (A) Average service time normalized to that of *"little + big opportunistic"*, (B) total CPU power, and (C) energy efficiency for the proposed and baseline scheduling polices.

only the little cluster results in the highest average service time per task, while additionally using the big cluster opportunistically when the little cluster is occupied reduces the average service time by almost 6×. An even further increase in performance by using the threshold policies, even though our proposed policies utilize the big cluster *less* than the policy that utilizing the big cores opportunistically. This is because the baseline policy over-utilizes big cores, generating thermal spikes that cause the system governor to throttle their frequency. Static in-order scheduling with task-based thresholds, the most constrained policy threshold policy, results in 1.4× reduction in average service time compared to little + big opportunistic. The best threshold policy (out-of-order scheduling with dynamic, time-based thresholds) reduces average service time by almost 4.8 × over little + big opportunistic.

To quantify the benefits of the proposed scheduling policies in meeting SLAs, Fig. 16 shows the cumulative distribution function (*cdf*) of service time for the different policies. Observe that for any deadline constraint, the best performing threshold policy significantly increases the fraction of jobs that

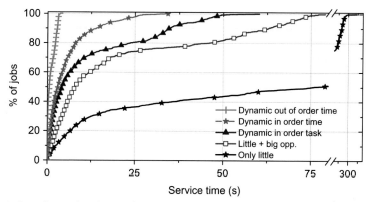

Fig. 16 Cumulative distribution function plot of the different configurations for service time.

meet the deadline compared to the two baseline policies. Compared to in-order scheduling policies, the *cdf* curve for out-of-order scheduling has a shorter tail, illustrating that it is indeed effective in ensuring that critical jobs meet their deadlines.

Compared to the lowest power policy that utilizes little cores only (4.44 W), threshold proposed policies increase the power consumption marginally to 6.23 W (Fig. 15B), while providing 22 × reduction in average service time. At the other extreme, utilizing little + big opportunistic consumes almost 7.33 W, yet has 4.82× lower performance than the proposed threshold policies. That is, the proposed threshold policies achieve both better performance and lower power than opportunistically using both clusters by utilizing the big core only when needed. Note that these results are for total platform power consumption—the trends for CPU-only power consumption are similar.

Finally, Fig. 15C plots the energy efficiency of each policy. The proposed policies threshold achieve substantially better energy efficiency than both extremes, i.e., using only the little cluster and using both clusters opportunistically.

Fig. 17B–D shows the frequency of the big cluster, system temperature, and queue occupancy of the scheduler queue as a function of time. The corresponding rate at which search queries arrive for processing in the time window for which data are plotted is shown in Fig. 17A.

Note that except the little + big opportunistic policy, the big core frequency of all other policies is as the nominal value of 2.2 GHz. As mentioned before, temperature spikes while executing the little + big opportunistic policy cause frequency throttling of the big core.

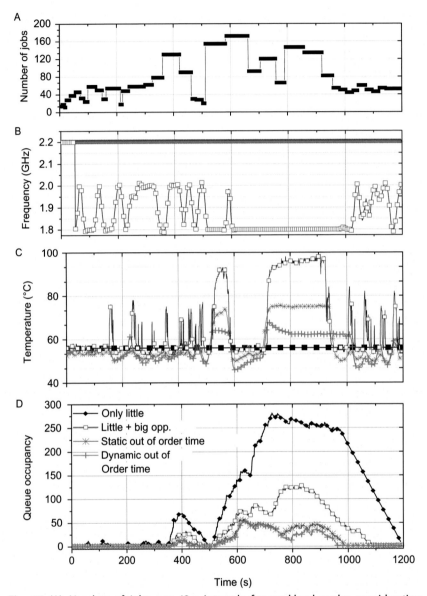

Fig. 17 (A) Number of jobs per 42-s intervals for workload under consideration. (B) Frequency of big core (A15). (C) Temperature of the big core (A15). (D) Queue occupancy for the different algorithms.

In contrast to the little + big opportunistic policy, the threshold policies run cooler with 10–30°C lower peak temperature, since they keep the big cores dark more often (see Fig. 18). The baseline policy that uses only little cores has the lowest peak temperature as expected. However, this comes at the expense of more than an order-of-magnitude performance loss, which can be explained by the large queue sizes that are observed when only little cores are used (see Fig. 17D).

Fig. 18 illustrates the activation of big cores with time for two policies, or equivalently, when the big cores are "dark." Observe that the threshold policy uses big cores less frequently than the little + big opportunistic policy, and yet provides better performance since it only uses big cores when absolutely required.

Table 3 highlights the top-line results from the empirical investigation of threshold scheduling policies for big.LITTLE processors. The key takeaway is the judicious use of the big core results in improvements in both performance and energy efficiency.

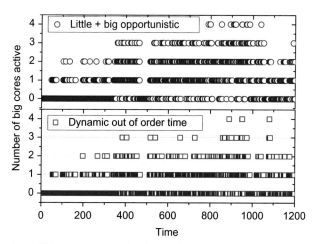

Fig. 18 Number of big cores active for "little + big opportunistic" and dynamic out of order time algorithms.

Table 3 Summary of the Improvements for the Different Scheduling Policies

Scheduling Policies	Static	Dynamic
In order, task	1.4×	1.53×
In order, time	1.8×	2.4×
Out of order, time	2.5×	4.8×

4. DYNAMIC DOP AND CLUSTER MIGRATION ON ASYMMETRIC MULTICORES

The ARM big.LITTLE processor architecture can be generalized to contain multiple clusters (or *pods*), where each pod contains cores of the same type. As an exemplar, in this section we will consider an asymmetric multicore, illustrated in Fig. 19, that has the following design:

- Multiple pods, where each pod consists of many identical processing cores and private caches. The core type used in a pod is different from the core type used in other pods on the chip—i.e., cores are micro-architecturally homogeneous within a pod but heterogeneous across pods. The cores in each pod support the same ISA [6]. Each pod consumes the full power budget of the chip (excluding uncore components), and thus, only one pod can be switched on at any given point in time while the others remain dark.

- A globally shared last level cache (LLC) with multiple banks and support for per-bank power gating [76]. This allows for the LLC cache capacity to be changed dynamically at run-time, i.e., at any given time, part of the LLC might be dark.

As in the previous section, assume that this processor is deployed in a datacenter to serve an incoming stream of service requests. Service requests (or *jobs*) are assumed to be independent (that is, they do not share state) and thus more requests can be served by scaling out to a larger number

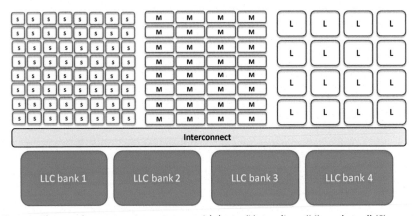

Fig. 19 Clustered asymmetric processor with large (L), medium (M), and small (S) cores. To meet the power constraint, only one cluster is allowed to be active at any time, while the other two remain dark.

of servers/cores. This is referred to as the scale-out software model [77]. Different from the assumptions in the previous section, we will assume that each request/job can be parallelized, and that the degree of parallelism (DoP) for each job can be determined at run-time.

A run-time scheduler makes the following decisions: (i) which pod to utilize, (ii) the DoP (D) of each job, (iii) the number of jobs to run in parallel (J) on that pod, and (iv) the number of banks of the LLC to turn on. For example, the scheduler can choose to execute jobs in sequence while parallelizing each job to the greatest possible degree. At the other extreme, the scheduler can execute multiple jobs in parallel, while assigning only one thread to each job. In general, assuming that the active pod has N_c cores, $J \times D = N_c$. The objective of the run-time scheduler is to minimize the mean service time of jobs within a peak power budget (recall that to meet the power constraint, only one pod can be turned on). As it turns out, the scheduler's optimal decisions depend on the job arrival rate λ. The design of one such scheduling policy, referred to as DoPpler Shift [3], is described here.

4.1 Related Work

Prior work on heterogeneous architectures has primarily focused on application heterogeneity as the reason for provisioning the chip with cores of many types [6, 78, 79]. Scheduling policies for clustered architectures, for example, on an architecture with one cluster of two Cortex A15 cores and another cluster of Cortex A7 cores, have been discussed to adapt to temporal variations in workload characteristics [80]. The empirical results discussed here show that as the optimal DoP and cluster/pod choice varies with varying job arrival rates, even if the application and workload characteristics do not vary.

Adaptive DoP at run-time has been proposed by prior researchers in the context of performance and power improvements [58, 81, 82]. However, the focus has primarily been on homogeneous architectures.

The idea of multiple "pods" on a chip was proposed recently in [77] but in an entirely different context. In this design, each pod is identical, and multiple pods can be turned on simultaneously. Also, the pods do not share any LLC and are disconnected from each other.

4.2 DoPpler Shift

For a setting in which jobs are constantly arriving to be serviced, an important metric of performance is total service time, i.e., the sum of time spent

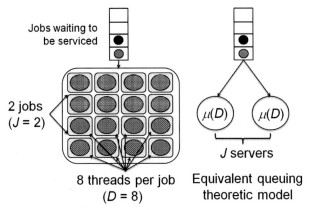

Fig. 20 A CMP with $N_c = 16$ cores. The scheduler schedules two jobs ($J = 2$) on the CMP, each with a DoP of three ($D = 8$). Also shown on the *right* is an equivalent queueing theoretic model of the system.

waiting in the queue and the execution time on the CMP. A queueing theoretic model can be used to express the mean total service time as a function of job arrival rate.

Assume a pod consisting of N_c homogeneous cores processing jobs that arrive at the server for processing at arrival rate λ according to a Poisson process. The DoP of each job is D and up to $J = \frac{N_c}{D}$ jobs can be processed in parallel. The mean service time, $\mu(D)$, for a job depends on the DoP D, and can be modeled as an M/M/n queue with $n = J$, as shown in Fig. 20. Now, using standard results, the mean total service time, W, of jobs can be written as follows:

$$W = \frac{p\left(J, \frac{\lambda}{\mu(D)}\right)}{J\mu(D) - \lambda} + \frac{1}{\mu(D)} \tag{1}$$

where $p\left(J, \frac{\lambda}{\mu(D)}\right)$ is the probability that a job has to wait in the queue to be serviced.

The expression for $p\left(J, \frac{\lambda}{\mu(D)}\right)$ is expressed in terms of the utilization of the server, ρ, which is written as:

$$\rho = \frac{\lambda}{J \times \mu(D)}, \tag{2}$$

resulting in the following expression for $p\left(J, \frac{\lambda}{\mu(D)}\right)$:

$$p\left(J, \frac{\lambda}{\mu(D)}\right) = \frac{\frac{(J \times \rho)^J}{J!} \times \frac{1}{1-\rho}}{\sum_{K=0}^{J-1} \frac{(J \times \rho)^K}{K!} + \frac{(J \times \rho)^J}{J!} \times \frac{1}{1-\rho}}. \tag{3}$$

The execution time of a program as a function of its DoP is governed by Amdahl's law [83], and hence,

$$\mu(D) = \frac{\mu_{seq}}{S + \frac{1-S}{D}} \tag{1}$$

where μ_{seq} is the service rate if the job is executed sequentially, and S is the fraction of execution that cannot be parallelized. S is also referred to as the serial fraction. This completes the analytical model. Several observations can be made from the model.

Remark 1. For $S = 0$, the optimal DoP, $D^* = N_c$, and $J^* = 1$. That is, perfectly parallelizable jobs should always be executed at highest DoP.

Remark 2. For $S > 0$ and $\lambda \to 0$, the optimal DoP, $D^* = N_c$ and $J^* = 1$. That is, when job arrival rate is low, jobs should be executed at the highest DoP.

Remark 3. For $S > 0$, the optimal DoP that sustains the highest arrival rate is $D^* = 1$, and $J^* = N_c$. That is, for a sufficiently high job arrival rate, jobs should be executed sequentially, with multiple jobs executing in parallel.

Fig. 21 shows the mean service time vs arrival rate for different DoPs with $S = 0.1$. It can be observed that the optimal DoP reduces with increasing arrival rates.

Now, to understand the impact of cluster/pod type on performance, assume an asymmetric multicore with T clusters/pods. The peak power consumption of a core in cluster $t \in T$ is given by P_c^t, and that the chip wide power budget (excluding uncore components) is given by P_{budget}. Thus, the number of active cores in cluster t, therefore, is

$$N_c^t = \left\lfloor \frac{P_{budget}}{P_c^t} \right\rfloor$$

Finally, assume that the mean sequential execution time of a job on a core of type t is μ_{seq}^t.

It can be shown that when $\lambda \to 0$, the service rate can be written as:

$$\frac{1}{W^t} = \mu(D = N_c^t) = P_{budget} \frac{\mu_{sec}^t}{SP_{budget} + (1-S)P_c^t},$$

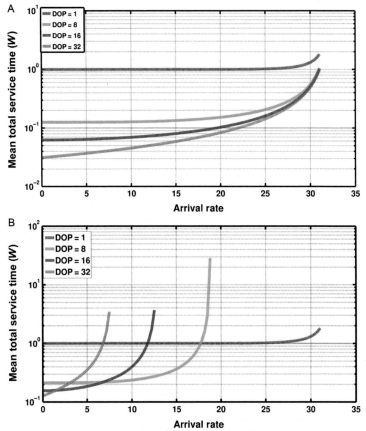

Fig. 21 Mean total service time as a function of arrival rate for different DoPs: (A) perfectly parallelizable jobs; (B) jobs with 10% serial fraction.

and conversely, for high arrival rates, the maximum arrival rate that cluster t $\in T$ can support is given by:

$$\lambda_{max}^{t} = P_{budget} \frac{\mu_{seq}^{t}}{P_{c}^{t}}$$

Based on these equations, we observe that the optimal cluster/pod choice for high arrival rates is the cluster/pod with the most power efficient cores. For low arrival rates, the optimal cluster choice depends on the application's serial fraction S. As S increases, the optimal cluster choice moves from slower, low power cores to toward clusters with faster, more power hungry cores.

DoPpler Shift's run-time scheduler assumes precharacterized information about the service rate for each pod type and DoP, i.e., $\mu^t(D)$, and uses Eq. (1) to pick the optimal DoP and cluster/pod given the current job arrival rate. Since the scheduler needs to make decisions into the future, it needs to predict the future arrival rate. A simple predictor that assumes that the arrival rate in the next control interval is the same as in the current control interval is used.

5. EMPIRICAL ANALYSIS

DoPler Shift is evaluated on a simulated architecture consisting of three pods (with 64, 32, and 16 cores each) and a 16 MB LLC (with four 4 MB banks) as shown in Fig. 19 at the 11 nm technology node. Table 4 details the microarchitectural parameters of the cores (small, medium, and large) in each of the three pods. The total chip power budget (TDP) is set to 115 W, of which 85 W is for the core components and the remaining for the uncore components. Note that each pod by itself consumes the entire 85 W core power budget.

Experimental results are obtained on the *Sniper* multicore simulator [48] using benchmark applications from the SPLASH-2 [36], PARSEC [84], and Phoenix [85] benchmark suites. More details about the benchmark applications used are in Table 5.

To determine the mean total service time, thousands of jobs need to be simulated, which can be prohibitively time-consuming on a detailed microarchitectural simulator. A hybrid approach that combines execution time data from the *Sniper* with a Python-based discrete event simulation (DES) engine based on SimPy[b] is therefore used. The hybrid simulator is referred

Table 4 Core Microarchitectural Details

Core Type	Nom. Freq. (GHz)	Dispatch Width	Window Size	Peak Power (W)	L1-D (KB)	L1-I (KB)	Area (×)
Small (S)	4.5	1	16	1.29	64	64	1
Medium (M)	4.5	2	128	2.59	64	64	1.33
Large (L)	4.5	4	128	5.18	128	128	1.66

[b] http://simpy.sourceforge.net/.

Table 5 Details of Benchmarks Used for Empirical Evaluation

Application	Description
Barnes	Simulation of interaction
Cholesky	Sparse matrix factorization
FFT	Fast fourier transform
Radix	Sorting
Raytrace	Image rendering
LU.ncont	Dense matrix factorization
Blackscholes	financial option pricing
Bodytrack	Image processing
PCA	Statistical machine learning
Word count	Text processing
String match	Text processing
K-means	Clustering
Linear regression	Statistical machine learning

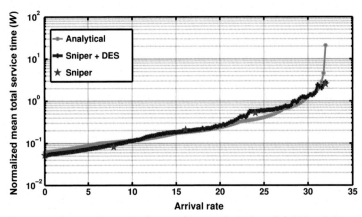

Fig. 22 Mean total service time predictions from analytical model, DES + Sniper and the Sniper multicore simulator for the Radix benchmark—medium core.

to as *Sniper* + DES. A similar simulation approach was proposed by [86]. Fig. 22 shows that the estimates of mean service time obtained from *Sniper* + DES are within 8.25% of *Sniper* over a wide range of arrival rates. All of the experiments reported in this section are based on *Sniper* + DES simulations. Fig. 22 also shows excellent agreement between the queueing theory-based

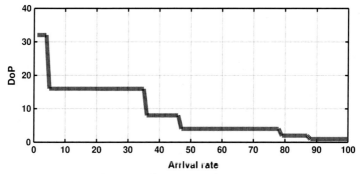

Fig. 23 Optimal DoP as a function of arrival rate for the FFT benchmark.

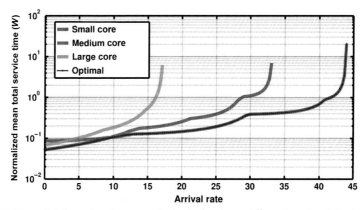

Fig. 24 Mean total service time vs arrival rate for three different pod architectures for the Radix benchmark.

analytical model for mean service time that DoPpler Shift uses and the data obtained from simulation.

The theoretical observations made regarding the dependence of optimal DoP and cluster/pod type on arrival rate are empirically validated in Fig. 23 (optimal DoP reduces with arrival rate) and Fig. 24 (optimal pod choice moves from pods with higher performance/higher power cores to more power efficient cores with increasing arrival rates). Further, Fig. 25 indicates that *dynamically* adapting the DoP to varying arrival rates outperforms a static scheduler that uses the best static DoP.

Typically, there is a discrepancy between the execution time of a single job (milliseconds to seconds) and the variation in arrival rate (over minutes or hours). In Fig. 26, the mean number of jobs that execute per update in either DoP or pod type is more than 20,000. Thus, the performance overhead of

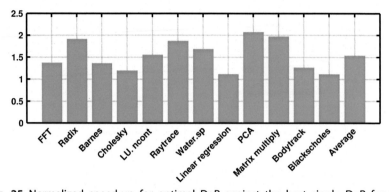

Fig. 25 Normalized speed-up for optimal DoP against the best single DoP for the medium core type under varying arrival rates.

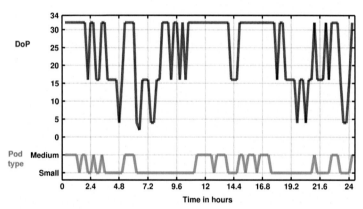

Fig. 26 Graph showing the variation in optimal DoP and core type for the Radix benchmark under varying arrival rate.

updating the DoP or switching to another pod is amortized over the relatively long period of time spent in that configuration. In the ARM big.little architecture it is seen that migration takes anywhere between 30 and 50 μs to transfer the architectural state between clusters. A conservative migration overhead of 50 μs is added when the pods need to be switched.

Fig. 26 shows a trace of the scheduler's decisions in response to job arrival rate variations. For this benchmark, the scheduler switches between the small and medium cluster/pod, while the large cluster/pod is left dark. (Other benchmarks do make use of the large cluster/pod.) Note also the frequent changes in DoP as the scheduler migrates jobs from one cluster/pod to another.

To understand the benefits of overprovisioning a processor with multiple clusters, DoPpler Shift is compared with

(a) conventional homogeneous multicore processors, with small, medium, and big cores only;

(b) a heterogeneous multicore processor with 0% dark silicon based on a modified version of the ARM big.LITTLE design; and

(c) a multicore processor that uses DVFS to meet the power budget.

5.1 Compared to Homogeneous Multicore + No Dark Silicon

The three conventional CMPs we compare with have 64 small cores, 32 medium cores and 16 large cores, respectively, and a 16 MB LLC. All architectures have the *same* peak power budget as DoPpler Shift but consume less area since they have no dark silicon. The mean service time obtained using DoPpler shift and using the three homogeneous multicore architectures is shown in Table 6. DoPpler Shift *reduces mean total service time*

Table 6 Tabulated Is the % Reduction in Mean Total Service Time Obtained Using DoPpler Shift With Respect to Homogeneous Multicore Architectures With Small, Medium, and Large Cores

	DoPpler Shift Speed-Up		
Application	% Over Small	% Over Medium	% Over Large
Barnes	75.68	18.18	43.75
Blackscholes	12.50	46.15	99.78
Bodytrack	36.00	15.79	97.09
Cholesky	79.32	7.55	86.27
FFT	55.21	4.44	66.92
Kmeans	20.37	0.00	94.45
Linear regression	22.67	44.23	98.62
LU.ncont	86.19	3.85	96.76
PCA	14.29	25.00	99.27
Radix	35.71	83.78	99.64
Raytrace	63.01	12.90	10.00
String match	13.64	61.03	99.24
Word count	1.33	83.11	99.50
Average	42.99	33.83	90.94

Small (medium, large) refers to a homogeneous multicore with 64 (32,16) small (medium, large) cores.

average over all benchmarks by between 29% and 90%, compared to conventional homogeneous multicore designs with no dark silicon.

5.2 Compared With Heterogeneous Multicore + No Dark Silicon

DoPpler Shift is compared with a heterogeneous multicore with 16 small cores, 8 medium core, and 4 large cores. This is similar in spirit to the heterogeneous platform discussed in [80], although it has three (instead of two) clusters. With *all* cores turned on, the architecture has the same peak power dissipation as DoPpler Shift, but has no dark silicon. Each arriving job is spawned with a DoP of four. The scheduler tries to schedule jobs to the *fastest* available set of cores, for example, when the pod with the four large cores is available, a job will be assigned to it first before being assigned to any other pod. DoPpler Shift is still 33.79% better in terms of mean total service time than the heterogeneous CMP with no dark silicon.

5.3 Compared With Homogeneous Multicore With DVFS

Some recent proposals have explored the idea of homogeneous, dark silicon CMPs using DVFS [87] (also called dim silicon), i.e., only a few cores turned on at a high voltage/frequency (V/F) level and more cores turned on at lower V/F levels. DoPpler Shift is compared with a homogeneous, dark silicon multicore with 96 medium cores and three available V/F levels. This architecture has an area that is only marginally smaller (5%) than DoPpler Shift, and the same power budget. The processor runs in three modes. There are either 32 cores active at the highest voltage/frequency levels (see Table 7), 64 cores at medium voltage/frequency, or 96 cores at lowest voltage/frequency. In each mode, the chip consumes the entire power budget. At the highest V/F level, only 32 cores can be turned on, while at medium (low) V/F level 64 (all 96) cores are turned on. It is empirically observed that DoPpler Shift is *11.64% better* in terms of mean total service time than the homogeneous dark silicon multicore (Fig. 27).

Table 7 V/F Pairs Used for the Homogeneous Dark Silicon Architecture

Voltage (V)	Frequency (GHz)
0.7	4.5
0.6	2.93
0.55	2.16

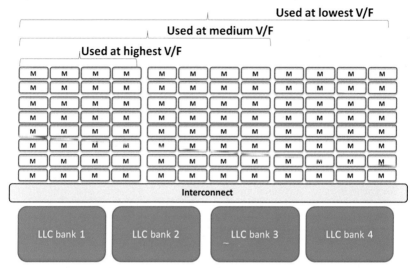

Fig. 27 Dim Silicon architecture consisting of 96 medium cores. Three modes of operation are possible: 32 cores active at 0.7 V, 64 cores active at 0.6 V, or all cores active at 0.55 V.

6. CONCLUSION

In this chapter, we have surveyed the state of the art on run-time power management techniques for over-provisioned homogeneous and asymmetric multicores. We have also discussed in greater detail three specific techniques [1–3] as examples of the types of solutions that need to be deployed to resolve the challenges posed by the presence of dark silicon.

REFERENCES

[1] Y. Turakhia, G. Liu, S. Garg, D. Marculescu, Thread progress equalization: dynamically adaptive power-constrained performance optimization of multi-threaded applications, IEEE Trans. Comput. 66 (4) (2017) 731–744.

[2] S. Jain, H. Navale, U. Ogras, S. Garg, Energy efficient scheduling for web search on heterogeneous microservers, in: 2015 IEEE/ACM International Symposium on Low Power Electronics and Design (ISLPED), IEEE, 2015, pp. 177–182.

[3] B. Raghunathan, S. Garg, Job arrival rate aware scheduling for asymmetric multi-core servers in the dark silicon era, in: Proceedings of the 2014 International Conference on Hardware/Software Codesign and System Synthesis, ACM, 2014, p. 14.

[4] H. Esmaeilzadeh, et al., Dark silicon and the end of multicore scaling, in: Proc. of Int. Symp. of Computer Architecture, IEEE, 2011, pp. 986–994.

[5] G. Venkatesh, J. Sampson, N. Goulding, S. Garcia, V. Bryksin, J. Lugo-Martinez, S. Swanson, M.B. Taylor, Conservation cores: reducing the energy of mature computations, in: ACM SIGARCH Computer Architecture News, vol. 38, ACM, 2010, pp. 205–218.

[6] R. Kumar, K.I. Farkas, et al., Single-ISA heterogeneous multi-core architectures: The potential for processor power reduction, in: MICRO-36, IEEE, 2003, pp. 81–92.

[7] T.S. Muthukaruppan, et al., Hierarchical power management for asymmetric multi-core in dark silicon era, in: Proc. of DAC, 2013.

[8] ARM, big.LITTLE Processing, http://www.arm.com/products/processors/technologies/biglittleprocessing.php.

[9] T.D. Burd, T.A. Pering, et al., A dynamic voltage scaled microprocessor system, IEEE J. Solid State Circuits 35 (11) (2000) 1571–1580.

[10] T.D. Burd, R.W. Brodersen, Design issues for dynamic voltage scaling, in: ISLPED, IEEE, 2000, pp. 9–14.

[11] G. Semeraro, G. Magklis, et al., Energy-efficient processor design using multiple clock domains with dynamic voltage and frequency scaling, in: HPCA, IEEE, 2002, pp. 29–40.

[12] A. Iyer, D. Marculescu, Run-time scaling of microarchitecture resources in a processor for energy savings, in: Cool Chips Workshop, Held in Conjunction with MICRO, vol. 33, 2000.

[13] P. Petrica, A.M. Izraelevitz, et al., Flicker: a dynamically adaptive architecture for power limited multicore systems, in: ISCA, ACM, 2013, pp. 13–23.

[14] S. Herbert, D. Marculescu, Analysis of dynamic voltage/frequency scaling in chip-multiprocessors, in: ISLPED, IEEE, 2007, pp. 38–43.

[15] C. Isci, A. Buyuktosunoglu, et al., An analysis of efficient multi-core global power management policies: maximizing performance for a given power budget, in: MICRO-39, IEEE, 2006, pp. 347–358.

[16] T. Ebi, M. Faruque, J. Henkel, Tape: thermal-aware agent-based power econom multi/many-core architectures, in: ICCAD, IEEE, 2009, pp. 302–309.

[17] J. Sartori, R. Kumar, Distributed peak power management for many-core architectures, in: DATE, IEEE, 2009, pp. 1556–1559.

[18] K. Ma, X. Li, M. Chen, X. Wang, Scalable power control for many-core architectures running multi-threaded applications, in: ACM SIGARCH Computer Architecture News, vol. 39, ACM, 2011, pp. 449–460.

[19] H. Hoffmann, S. Sidiroglou, et al., Dynamic knobs for responsive power-aware computing, in: ACM SIGPLAN Notices, vol. 46, ACM, 2011, pp. 199–212.

[20] R. Cochran, C. Hankendi, et al., Pack & Cap: adaptive DVFS and thread packing under power caps, in: MICRO, ACM, 2011, pp. 175–185.

[21] W. Godycki, C. Torng, I. Bukreyev, A. Apsel, C. Batten, Enabling realistic fine-grain voltage scaling with reconfigurable power distribution networks, in: Proceedings of the 47th Annual IEEE/ACM International Symposium on Microarchitecture, IEEE Computer Society, 2014, pp. 381–393.

[22] K.K. Rangan, G.-Y. Wei, D. Brooks, Thread motion: fine-grained power management for multi-core systems, in: ACM SIGARCH Computer Architecture News, vol. 37, ACM, 2009, pp. 302–313.

[23] N.B. Lakshminarayana, J. Lee, et al., Age based scheduling for asymmetric multiprocessors, in: Proceedings of the Conference on High Performance Computing Networking, Storage and Analysis, ACM, 2009, p. 25.

[24] K. Du Bois, S. Eyerman, et al., Criticality stacks: identifying critical threads in parallel programs using synchronization behavior, in: ISCA, ACM, 2013, pp. 511–522.

[25] J.A. Joao, M.A. Suleman, et al., Bottleneck identification and scheduling in multi-threaded applications, in: ASPLOS, ACM, 2012, pp. 223–234.

[26] J.A. Joao, M.A. Suleman, et al., Utility-based acceleration of multithreaded applications on asymmetric CMPs, in: ISCA, ACM, 2013, pp. 154–165.

[27] A. Bhattacharjee, M. Martonosi, Thread criticality predictors for dynamic performance, power, and resource management in chip multiprocessors, in: ACM SIGARCH Computer Architecture News, vol. 37, ACM, 2009, pp. 290–301.

[28] K. Van Craeynest, S. Akram, et al., Fairness-aware scheduling on single-ISA heterogeneous multi-cores, in: PACT, IEEE, 2013, pp. 177–187.

[29] M.C. Huang, J. Renau, J. Torrellas, Positional adaptation of processors: application to energy reduction, in: 30th Annual International Symposium on Computer Architecture, 2003, IEEE, 2003, pp. 157–168.

[30] D. Ponomarev, G. Kucuk, K. Ghose, Reducing power requirements of instruction scheduling through dynamic allocation of multiple datapath resources, in: Proceedings of the 34th Annual ACM/IEEE International Symposium on Microarchitecture, IEEE Computer Society, 2001, pp. 90–101.

[31] A. Buyuktosunoglu, T. Karkhanis, et al., Energy efficient co-adaptive instruction fetch and issue, in: ISCA, IEEE, 2003, pp. 147–156.

[32] H.R. Ghasemi, N.S. Kim, RCS: runtime resource and core scaling for power-constrained multi-core processors, in: Proceedings of PACT, ACM, 2014, pp. 251–262.

[33] E. Ipek, M. Kirman, et al., Core fusion: accommodating software diversity in chip multiprocessors, ACM SIGARCH Computer Architecture News 35 (2) (2007) 186–197.

[34] R. Rodrigues, A. Annamalai, et al., Performance per watt benefits of dynamic core morphing in asymmetric multicores, in: PACT, IEEE, 2011, pp. 121–130.

[35] G. Liu, J. Park, D. Marculescu, Dynamic thread mapping for high-performance, power-efficient heterogeneous many-core systems, in: ICCD, IEEE, 2013, pp. 54–61.

[36] S.C. Woo, M. Ohara, et al., The SPLASH-2 programs: characterization and methodological considerations, in: ISCA, ACM, 1995, pp. 24–36.

[37] E.Z. Zhang, Y. Jiang, X. Shen, Does cache sharing on modern CMP matter to the performance of contemporary multithreaded programs? in: ACM Sigplan Notices, vol. 45, ACM, 2010, pp. 203–212.

[38] J.P. Singh, C. Holt, et al., Load balancing and data locality in adaptive hierarchical N-body methods: Barnes-Hut, fast multipole, and radiosity. J. Parallel Distrib. Comput. 27 (2) (1995) 118–141. ISSN: 0743-7315, https://doi.org/10.1006/jpdc.1995.1077.

[39] C. Liu, A. Sivasubramaniam, et al., Exploiting barriers to optimize power consumption of CMPs, in: 19th IEEE International Parallel and Distributed Processing Symposium, 2005. Proceedings, IEEE, 2005, p. 5a.

[40] E. Duesterwald, J. Torrellas, S. Dwarkadas, Characterizing and predicting program behavior and its variability, in: PACT, 2003, pp. 220–231.

[41] A. Lukefahr, S. Padmanabha, et al., Composite cores: pushing heterogeneity into a core, in: MICRO, IEEE Computer Society, 2012, pp. 317–328.

[42] K. Van Craeynest, A. Jaleel, et al., Scheduling heterogeneous multi-cores through performance impact estimation (PIE), in: ISCA, ACM, 2012, pp. 213–224.

[43] B. Sprunt, Pentium 4 performance-monitoring features, Micro, IEEE 22 (4) (2002) 72–82.

[44] Q. Liang, IBM, Performance Monitor Counter Data analysis Using Counter Analyzer (Online). 2009, www.ibm.com/developerworks/aix/library/au-counteranalyzer/.

[45] S. Eyerman, L. Eeckhout, et al., A performance counter architecture for computing accurate CPI components, in: ASPLOS, ACM, 2006, pp. 175–184.

[46] G. Contreras, M. Martonosi, Power prediction for intel XScale® processors using performance monitoring unit events, in: ISLPED, IEEE, 2005, pp. 221–226.

[47] W.L. Bircher, L.K. John, Complete system power estimation using processor performance events, IEEE Trans. Commun. 61 (4) (2012) 563–577.

[48] T.E. Carlson, W. Heirman, L. Eeckhout, Sniper: exploring the level of abstraction for scalable and accurate parallel multi-core simulation, in: Proceedings of 2011 International Conference for High Performance Computing, Networking, Storage and Analysis, ACM, 2011, p. 52.
[49] ODROID–XU3, http://www.hardkernel.com/main/main.php.
[50] W. Lin, P.R. Kumar, Optimal control of a queueing system with two heterogeneous servers, IEEE Trans. Autom. Control 29 (8) (1984) 696–703.
[51] S. Jin, G. Schiavone, D. Turgut, A performance study of multiprocessor task scheduling algorithms, J. Supercomput. 43 (1) (2008) 77–97.
[52] Y.-K. Kwok, I. Ahmad, Static scheduling algorithms for allocating directed task graphs to multiprocessors, ACM Comput. Surv. (CSUR) 31 (4) (1999) 406–471.
[53] T.D. Braun, H.J. Siegal, N. Beck, L.L. Boloni, M. Maheswaran, A.I. Reuther, J.P. Robertson, M.D. Theys, B. Yao, D. Hensgen, et al., A comparison study of static mapping heuristics for a class of meta-tasks on heterogeneous computing systems, in: Eighth Heterogeneous Computing Workshop, 1999 (HCW'99). Proceedings, IEEE, 1999, pp. 15–29.
[54] O.H. Ibarra, C.E. Kim, Heuristic algorithms for scheduling independent tasks on non-identical processors, J. ACM (JACM) 24 (2) (1977) 280–289.
[55] L. Wenjing, W. Lisheng, Energy-considered scheduling algorithm based on heterogeneous multi-core processor, in: 2011 International Conference on Mechatronic Science, Electric Engineering and Computer (MEC), IEEE, 2011, pp. 1151–1154.
[56] M. Becchi, P. Crowley, Dynamic thread assignment on heterogeneous multiprocessor architectures, in: Proc. of Conference on Computing Frontiers, 2006, pp. 29–40.
[57] D. Koufaty, D. Reddy, S. Hahn, Bias scheduling in heterogeneous multi-core architectures, in: Proceedings of the 5th European conference on Computer systems, ACM, 2010, pp. 125–138.
[58] D. Shelepov, J.C. Saez Alcaide, et al., HASS: a scheduler for heterogeneous multicore systems, ACM SIGOPS Oper. Syst. Rev. 43 (2) (2009) 66–75.
[59] V. Gupta, R. Nathuji, Analyzing performance asymmetric multicore processors for latency sensitive datacenter applications, in: Proceedings of the 2010 International Conference on Power Aware Computing and Systems, USENIX Association, 2010, pp. 1–8.
[60] B. Raghunathan, S. Garg, Job arrival rate aware scheduling for asymmetric multi-core servers in the dark silicon era, in: Proc. of CODES, 2014.
[61] Y. Zhu, V.J. Reddi, High-performance and energy-efficient mobile web browsing on big/little systems, in: Proc. of Int. Symp. on HPCA, 2013, pp. 13–24.
[62] T. Heath, B. Diniz, E.V. Carrera, W. Meira Jr., R. Bianchini, Energy conservation in heterogeneous server clusters, in: Proceedings of the Tenth ACM SIGPLAN Symposium on Principles and Practice of Parallel Programming, ACM, 2005, pp. 186–195.
[63] M. Pricopi, et al., Power-performance modeling on asymmetric multi-cores, in: Proc. of CASES, 2013, pp. 1–10.
[64] J. Walrand, A note on "optimal control of a queuing system with two heterogeneous servers" Syst. Control Lett. 4 (3) (1984) 131–134.
[65] G. Koole, A simple proof of the optimality of a threshold policy in a two-server queueing system, Syst. Cont. Lett. 26 (5) (1995) 301–303.
[66] N.M. Asghari, M. Mandjes, A. Walid, Energy-efficient scheduling in multi-core servers, Comput. Netw. 59 (2014) 33–43.
[67] V. Janapa Reddi, et al., Web search using mobile cores: quantifying and mitigating the price of efficiency, in: ACM SIGARCH Computer Arch. News, 2010, pp. 314–325.
[68] M. Guevara, et al., Navigating heterogeneous processors with market mechanisms, in: Proc. of High Performance Computer Architecture, 2013, pp. 95–106.
[69] C. Delimitrou, C. Kozyrakis, Paragon: QoS-aware scheduling for heterogeneous datacenters, in: Proc. of ASPLOS, 2013, 2013, pp. 77–88.

[70] G. Wu, Z. Xu, Q. Xia, J. Ren, F. Xia, Task allocation and migration algorithm for temperature-constrained real-time multi-core systems, in: Proceedings of the 2010 IEEE/ACM Int'l Conference on Green Computing and Communications & Int'l Conference on Cyber, Physical and Social Computing, IEEE Computer Society, 2010, pp. 189–196.

[71] Y. Xie, W.-L. Hung, Temperature-aware task allocation and scheduling for embedded multiprocessor systems-on-chip (MPSoC) design, J. VLSI Signal Process. Syst. Signal Image Video Technol. 45 (3) (2006) 177–189.

[72] Apache Software Foundation, Lucene Search Engine, http://www.ibm.com/developerworks/library/os-apache-lucenesearch/.

[73] Y. Chen, A. Ganapathi, R. Griffith, R. Katz, The case for evaluating MapReduce performance using workload suites, in: 2011 IEEE 19th International Symposium on Modeling, Analysis & Simulation of Computer and Telecommunication Systems (MASCOTS), IEEE, 2011, pp. 390–399.

[74] E. Bragg, M. Guevara, B.C. Lee, Understanding query complexity and its implications for energy-efficient web search, in: Proceedings of the International Symposium on Low Power Electronics and Design, IEEE Press, 2013, pp. 401–401.

[75] Apache Software Foundation, Websites Powered by Lucene Search Engine, http://wiki.apache.org/lucene-java/PoweredBy.

[76] S. Ramaswamy, S. Yalamanchili, Improving cache efficiency via resizing+ remapping, in: 25th International Conference on Computer Design, 2007: ICCD 2007, IEEE, 2007, pp. 47–54.

[77] P. Lotfi-Kamran, B. Grot, M. Ferdman, S. Volos, O. Kocberber, J. Picorel, A. Adileh, D. Jevdjic, S. Idgunji, E. Ozer, et al., Scale-out processors, in: ACM SIGARCH Computer Architecture News, vol. 40, IEEE Computer Society, 2012, pp. 500–511.

[78] H. Esmaeilzadeh, E. Blem, et al., Dark silicon and the end of multicore scaling, in: 2011 38th Annual International Symposium on Computer Architecture (ISCA), IEEE, 2011, pp. 365–376.

[79] Y. Turakhia, B. Raghunathan, S. Garg, D. Marculescu, HaDeS: architectural synthesis for heterogeneous dark silicon chip multi-processors, in: Proceedings of the 50th Annual Design Automation Conference, ACM, 2013, p. 173.

[80] T.S. Muthukaruppan, M. Pricopi, V. Venkataramani, T. Mitra, S. Vishin, Hierarchical power management for asymmetric multi-core in dark silicon era, in: 2013 50th ACM/EDAC/IEEE Design Automation Conference (DAC), IEEE, 2013, pp. 1–9.

[81] J. Li, J.F. Martinez, Dynamic power-performance adaptation of parallel computation on chip multiprocessors, in: The Twelfth International Symposium on High-Performance Computer Architecture, 2006, IEEE, 2006, pp. 77–87.

[82] A. Raman, H. Kim, T. Oh, J.W. Lee, D.I. August, Parallelism orchestration using DoPE: the degree of parallelism executive, in: ACM SIGPLAN Notices, vol. 46, ACM, 2011, pp. 26–37.

[83] D.H. Woo, H.-H.S. Lee, Extending Amdahl's law for energy-efficient computing in the many-core era, Computer 41 (12) (2008).

[84] C. Bienia, S. Kumar, et al., The PARSEC benchmark suite: characterization and architectural implications, in: PACT, ACM, 2008, pp. 72–81.

[85] C. Ranger, R. Raghuraman, et al., Evaluating mapreduce for multi-core and multiprocessor systems, in: HPCA, IEEE, 2007, pp. 13–24.

[86] D. Meisner, J. Wu, T.F. Wenisch, Bighouse: a simulation infrastructure for data center systems, in: 2012 IEEE International Symposium on Performance Analysis of Systems and Software (ISPASS), IEEE, 2012, pp. 35–45.

[87] M.B. Taylor, Is dark silicon useful?: Harnessing the four horsemen of the coming dark silicon apocalypse, in: DAC, ACM, 2012, pp. 1131–1136.

ABOUT THE AUTHOR

Siddharth Garg received his Ph.D. degree in Electrical and Computer Engineering from Carnegie Mellon University in 2009, and a B.Tech. degree in Electrical Engineering from the Indian Institute of Technology Madras. He joined NYU in Fall 2014 as an Assistant Professor, and prior to that, was an Assistant Professor at the University of Waterloo from 2010–2014. His general research interests are in computer engineering, and more particularly in secure, reliable and energy-efficient computing.

CHAPTER SIX

Topology Specialization for Networks-on-Chip in the Dark Silicon Era

Mehdi Modarressi*,†,1, Hamid Sarbazi-Azad†,‡
*College of Engineering, University of Tehran, Tehran, Iran
†School of Computer Science, Institute for Research in Fundamental Sciences (IPM), Tehran, Iran
‡Sharif University of Technology, Tehran, Iran
1Corresponding author: e-mail address: modarressi@ut.ac.ir

Contents

Advances in Computers, Volume 110
ISSN 0065-2458
https://doi.org/10.1016/bs.adcom.2018.03.009

Abstract

Following Moore's law, the number of transistors on chip has grown exponentially for decades. This growing transistor count, coupled with recent architecture and compiler advances, has resulted in an unprecedented exponential performance increase of computers. With the end of Dennard scaling, however, the power required to operate all transistors at the full performance level simultaneously grows across the technology generations. Consequently, chips will keep an increasing fraction of transistors power gated or dark to remain within the power envelope. The power-gated part of the chip, known as *dark silicon*, is expected to comprise a significant portion of the die real estate in new technology generations. In addition to power limitations, the limited and nonscalable off-chip bandwidth is the second source of dark silicon. The key challenge to improving performance in the dark silicon era, consequently, is how to efficiently leverage transistors when they cannot all be powered at the same time. Core specialization shows promise in addressing this challenge by improving the performance and energy efficiency of applications. This method leverages the dark silicon to build many diverse power-efficient accelerators and each application only activates a subset of cores that best match its processing requirements. In this chapter, we propose a network-on-chip specialization method that leverages dark routers of a partially active many-core chip to customize the topology for active cores, effectively reducing the power consumption and latency of communication.

1. INTRODUCTION

For decades, semiconductor technology scaling has been served as the main driver for the design and fabrication of devices with more, smaller, and faster transistors in each new technology generation [1]. As a result of this technology scaling, major semiconductor manufacturers already ship chips with 24 [2], 28 [3], 32 [4], and 72 cores [5], and many-core chip multiprocessor (CMP) with several hundreds to thousands of cores are likely to appear in the future [6–8]. Following the same trend, future system–on–chip implementations will contain hundreds of heterogeneous cores with different sizes and functionalities [7,9] that form a many-core multiprocessor system–on–chip (MPSoC). Although this large number of cores can potentially give a huge processing performance, power constraints, referred to as the power wall, limit the maximum achievable performance [8,10,11]. Due to this constraint, which mainly stems from the failure of Dennard scaling [10] (that will be elaborated shortly), increasing the core count beyond a limit can no longer be translated to performance improvements, because we do not have enough electrical power for all on-chip components to operate at the maximum frequency simultaneously.

In addition to the power wall, bandwidth wall is another limiting factor for the performance of current and future microprocessors [8]. The bandwidth wall limits the off-chip memory bandwidth, mainly due to the fact that the chip physical dimensions and the number of I/O pads cannot be scaled proportional to the transistor scaling rate with Moore's law, rather tend to remain largely unchanged when technology advances [8].

To stay within the limited power and bandwidth budgets, only a small fraction of on-chip transistors in future CMPs and MPSoCs can be turned on simultaneously, leaving the remaining fraction, referred to as *dark silicon*, power gated [8,10,11]. It has been shown that without architecture- and circuit-level improvements to overcome the power and bandwidth constraints, the dark silicon will comprise up to 90% of the chip area in future technology generations [10,12].

As progressively smaller portions of a chip's real estate can be fully utilized, silicon area becomes a cheaper resource relative to power and energy consumption [8]. This paradigm shift highlights the need for new architectural design methods that trade the area for energy and bandwidth efficiency. 3D memory-on-logic architectures and the processing-in-memory concept are the main solutions to mitigate bandwidth constraints [8,13–15]. Core specialization, which trades off the silicon area with energy efficiency, is considered as the most promising solution that targets the power wall [8,12,16]. This approach integrates a selection of diverse customized cores (accelerators) into a single many-core processor. Each application then activates and runs on those cores that best match its processing requirements. Such accelerator-rich architectures are envisioned to become commonplace in a variety of computing domains in near future [17].

Since active cores in a partially dark accelerator-rich CMP may not always form a contiguous active region, such partially active many-core CMPs require the special on-chip communication support to mitigate the performance loss that the long on-chip latency may impose. In this chapter, after elaborating on the dark silicon problem and discussing the best-known solutions to tackle the bandwidth and power walls, we propose to specialize the communication part of a many-core accelerator-rich chip in the dark silicon era.

Network-on-chip (NoC) is the most scalable and efficient communication mechanism in modern multi- and many-core systems [18,19]. The performance of an NoC is extremely sensitive to its topology, since topology determines the cost, diameter, average message hop count, and bisection bandwidth of a network. Finding a suitable topology for a multitask parallel application, as well as task to core mapping, to reduce the number of intermediate routers between communicating cores is critical for total system performance.

Several prior works explore the benefits of topology reconfiguration to optimize NoC power/performance parameters [20–23]. In Ref. [20] a reconfigurable NoC is presented that can dynamically adapt its topology to the communication pattern of a given application. The reconfiguration of this NoC is achieved by embedding simple FPGA-like switch boxes in the network to dynamically change the NoC internode connections. The switch boxes are used to build a shortcut path between nodes with high communication demand to reduce their topological distance and carry their packets with lower latency and power consumption. By customizing the topology for a given on-chip traffic pattern, this reconfigurable NoC can offer up to 30% reduction in power consumption and 35% reduction in latency, while imposing 10%–45% area overhead due to the extra links and switch boxes it requires [20].

In this chapter, we propose a reconfigurable NoC based on the architecture presented in Ref. [20] that leverages inactive routers of a many-core chip to customize the topology for active cores. In this design, routers of the dark part of the chip act as bypass switches and are leveraged to make a topology with low average hop count for active cores. Our experimental results show considerable reduction in NoC energy consumption and latency [24].

2. DARK SILICON

According to Moor's law, transistor density and speed continue to improve by $2\times$ and $1.4\times$ every 2 years [11], respectively. We define the scaling factor of two process generations (s) as the ratio of the feature sizes of them. Following the Moore's law, the semiconductor industry has been successfully keeping the scaling factor of $1.4\times$ for two consecutive process generations. This can be verified by considering the ratio between the feature size of the processes in recent years (90, 65, 45, 32, 22, 14, 11, and 7 nm).

For each generation, as both dimensions of a chip shrink by a factor of s, the available transistors on a chip increase by s^2, when the chip area is kept fixed. On the other hand, the capacitance and resistance of the transistors are totally reduced by $1/s$, effectively increasing native frequency of the transistors by s. Considering a fixed die area, these factors altogether result in a net $s^3 x$ improvement in computational power. With $s = 1.4$, we can expect to have $2.8\times$ improvement in performance per process generation.

This performance improvement comes at the price of higher power consumption. Recall that the dynamic power consumption of a chip is calculated as $P_{dynamic} = \alpha CV^2 f$. According to this relation, the power consumption is increased by s^3, as the number of transistors and the native frequency scale up by s^2 and s, respectively, in each new process generation. To maintain a constant power envelope, this s^3 potential power increase must be mitigated by a corresponding reduction in transistor dynamic power. The first reduction comes from scaling down the transistor size that makes the switching capacitance of transistors drop by a factor of s. Another s^2 power reduction is obtained by scaling down the operating voltage (and the threshold voltage) by a factor of s according to Dennard's scaling law [10,11].

As such, the dynamic power consumption of a transistor transition drops by s^3 (see the dynamic power relation), exactly matching the additional power consumption imposed by scaling. As a result, with Dennard voltage scaling, we were often able to operate all transistors at full speed.

However, the Dennard scaling started to break down in 2005 [16]. The main reason is that scaling down the working voltage involves in reducing the threshold voltage of the transistor proportionally in order to maintain transistor performance. In today's so-called post-Dennard scaling, however, scaling down the threshold voltage will increase the leakage power exponentially. This leakage power is so large in recent process generations that can nullify the power reduction caused by voltage scaling. Thus, the threshold voltage could not be scaled down anymore and will remain largely unchanged across next technology generations. On the other hand, chip power delivery and cooling limitations do not scale well and remain mostly unchanged as the technology advances [25].

This way, with the failure of traditional circuit-level mechanisms, architecture-level mechanisms are needed to reduce the power consumption by a factor of s^2, in order to keep the power usage constant when the feature size shrinks. More details about the Dennardian and post-Dennardian transistor scaling can be found in Refs. [11,12].

This extra power consumption, coupled with limitations in power delivery and heat dissipation systems, limits the peak achievable performance of modern processors. To stay within the power envelop, designers can either keep a large portion of chip power gated at any time interval (the dark silicon approach) [3] or operate all transistors, but at a lower voltage and frequency than the maximum (the dim silicon approach) [26,27].

Although the simplest approach seems to be building only the number of transistors that we can afford to power, this approach results in smaller chip area, thereby rising the high power density and off-chip bandwidth limitation (due to fewer I/O pads) again [12].

2.1 The Dim Silicon Approach

Dim silicon techniques have been widely adopted in today's multicore processors. The simplest and most common dimming technique is to set the working frequency to a value significantly lower than the maximum frequency that the processor can work in. This ensures that the processor will be able to handle essentially all applications without exceeding its predefined TDP (thermal design power). TDP is the nominal amount of power (or the equivalent heat) that the cooling system of a chip is designed to dissipate [28].

In addition to the permanent frequency scale down, dynamic voltage and frequency scaling and dynamic power management (through power gating and clock gating) methods have been considered by both industry and academia as efficient ways to cope with the power constraints. Recent studies show that the power wall can be pushed significantly by performance-aware scaling down the voltage and frequency levels [11,29]. Several previous work then try to optimize the voltage/frequency of each individual core (or group of cores) of a many-core processor based on a given power and performance constraint [29–32].

Dynamic frequency and voltage scaling have also been a widely used technique in commercial processors to make a dynamic trade-off between power and performance for each operating situation [33–37]. An example is the TurboCore technology implemented in modern AMD processors [35]. In a six-core AMD Phenom processor, this TDP-aware performance boosting technology automatically switches from six cores to three turbocharged cores, when running applications that need speed over a few cores. In TurboCore mode, the processor shifts frequency speed from 3.2 GHz on six cores, to 3.6 GHz on three cores. However, to stay within its TDP, TurboCore sets down the frequency of the other three cores to 800 MHz and drops their voltage. Another example is IBM Power8 architecture that supports per-core DVFS control with 69 distinct frequency levels [36].

2.2 The Dark Silicon Approach

The dark silicon approach, on the other hand, operates the transistors at higher frequencies (or ideally the highest frequency that the circuit allows), but powers off part of the transistors to stay within the TDP envelop. Many researchers have pursued different techniques to leverage the dark silicon to increase performance and power efficiency [10,12,16,38–40]. Industry has also adopted some simple dark silicon-based architectures in recent years [41,42].

ARM big.LITTLE is the first attempt to leverage dark silicon in a commercial product [41]. This architecture consists of four big ARM Cortex-A15 cores and four LITTLE Cortex-A7 cores. Each Cortex-A15 is paired with a Cortex-A7 to create a system that can accomplish both high compute-intensive and low compute-intensive tasks in an energy efficient manner. In the task-migration use model of this architecture, only one core from each pair and never both cores are powered on at the same time. At run-time, when Cortex-A7 reaches its highest operating point and more performance is still required, a task-migration process is invoked that activates Cortex-A15 and moves the application to it. Since all the cores have access to the same memory regions and feature the same ISA, workloads can be swapped between the big and the little cores on the fly. It has been shown that this architecture offers up to 75% power reduction in low to moderate performance scenarios [41].

Another example is the 4 + 1 architecture used in the nVidia Tegra K1 system-on-chip family [42]. This architecture consists of four "big" Cortex cores and one "battery saver" Cortex core. All five CPU cores are identical, but the battery saver core is built using a special low power silicon process to consume very low power, but at the price of very low performance. Big cores are built using a standard silicon process to reach higher frequencies. The 1 + 4 cores will never be powered on simultaneously, but at any time, only one (the battery saver) or four (big cores) cores are active, leaving the rest dark. Battery saver executes tasks at low frequency when the device is in an active standby mode and there is no demand for performance, whereas big cores are used to run applications when the device is active.

Advanced liquid cooling techniques can effectively decrease the chip temperature and push the power envelope much further than is possible with conventional air cooling systems [43]. For such systems, the dark silicon problem appears again due to the limited amount of power pins on the chip package. Several previous works tackle this problem by dynamically switching a portion of I/O pins to power pins during the time intervals with low off-chip bandwidth demand [44,45]. While these methods may put more pressure on the already limited memory bandwidth, a recent paper proposed mutual switching of pins between power and I/O mode [45]. This pin switching scheme is motivated by the observation that a processor requires less power during memory-intensive stages than is available. For such memory-intensive execution phases, the power delivery pins are converted to I/O to provide extra bandwidth. However, these pin switching methods may fall short in many-core processor when each core is in a different execution phase: some cores are in compute-intensive execution phases and some in memory-intensive phases, so there is demand for both power and bandwidth at the same time.

2.3 Bandwidth Wall

The limited and nonscalable off-chip bandwidth is the second source of dark silicon. In a conventional system with off-chip memory, the maximum available memory bandwidth is calculated on the basis of the number of pads and the off-chip clock frequency. Based on the ITRS and other technology projection reports [10,46], two-thirds of the pads in a typical many-core processor are used for power and ground, leaving at most one-third available for off-chip memory and I/O signaling.

Traditional approaches to alleviate off-chip bandwidth problem use larger LLC to remain within the bandwidth envelope or increase the off-chip clock frequency to push the bandwidth wall. However, these methods further fuel the power wall problem [10].

Not only the communication with off-chip memory limits the achievable performance but also has a significant and increasing contribution to the total system power consumption [47,48]: a 64-bit DRAM access (memory access plus data transfer between processor and off-chip memory) in recent technologies consumes nearly two orders of magnitude more energy than a double-precision floating-point arithmetic operation [48].

The emerging 3D integration technology shows promise in increasing the bandwidth and reducing the power/latency of memory access. Thanks to the recent progress in the semiconductor integration technology, 3D stacking now allows DRAM and logic dies, each implemented by a different semiconductor process, to be stacked on top of each other in a 3D architecture within a single package [49,50].

Through-silicon via (TSV) is the dominant technology to implement vertical interdie connections in a 3D architecture [51]. Due to shorter length and lower capacitance of TSVs, when compared to horizontal 2D on-chip and off-chip wires, data communication along the third dimension involves by far lower latency and energy consumption. The ultra-low latency of TSVs (on the order of a few picoseconds), coupled with the higher density of them, provides a substantial increase in bandwidth at lower energy relative to a conventional off-chip memory system [48,49]. Prior work shows that 3D memory stacking leads to significant performance improvement and pushes the bandwidth constraint beyond the power limits [10,49,50].

2.4 Processing in Memory

Thermal challenges, however, are a key impediment to stacking memory directly on top of a high-performance processor [48,52]. In 3D architectures, the mean temperature of the layers farther from the heat sink is higher due to the longer heat conduction path to the cooling system. The high power consumption of a high-performance processor increases heat conductivity and makes the thermal management more complicated. Consequently, the temperature of memory layers, which are often farther from the heat sink, increases. High temperature degrades the retention time of the 3D-stacked DRAM and makes them more susceptible to temperature-induced performance and reliability problems [52]. To address this problem, researchers propose to use simple power-efficient processing-in-memory (PIM) units instead of full-fledged high-performance processors in the logic layer of a 3D memory architecture [48,53–56]. The PIM units may be simple power-efficient in-order cores [48,54,55] or specialized accelerators [56]. In this design, the host processor and 3D memory (and memory-side processing units) are connected by an interposer carrier to form a 2.5D-stacked system [57]. In this scheme, the host processor offloads the memory-intensive parts of the code to the memory-side processing units to benefit from the excessive bandwidth provided by the 3D TSVs at the memory side.

Recent advances have made 2.5D-stacked DRAM architectures, such as HBM [58] and HMC [59], commercially viable. The memory-side architecture in such systems consists of several banked DRAM layers and a logic layer, where architects can implement memory controller and other logic that interacts with both the host processor and the DRAM layers [60]. PIM research proposals add PIM units (accelerators, simple cores) to this logic layer to exploit the high bandwidth available. The benefits of stacked memory and PIM are twofold: first the abundant power consumption of the processor-memory communication is reduced and second, by executing some parts of the code in memory, part of the off-chip bandwidth demand is eliminated. These effects allow processors to power on more cores with the same power and bandwidth envelop.

3. CORE SPECIALIZATION

Among different approaches, hardware and software researchers in industry and academia now focus on core specialization as the best way to leverage dark silicon for energy efficiency [10,16,39]. Core specialization trades the now cheaper area with power efficiency to keep more cores powered on with the same power budget. With specialized cores (which are also known as coprocessors or accelerators), the dark silicon is leveraged to build and integrate a selection of diverse application-specific cores into a single many-core chip.

Specialized cores are designed by characterizing the target workloads by run-time profiling or static analysis and then identifying the compute-intensive parts of the codes [10,16]. Afterward, these parts are implemented as a hardware functional unit to carry out the functionality in a performance- and power-efficient manner. Specialized cores may be implemented as GPUs, DSPs, and FPGAs, but ASIC accelerators, which implement algorithms in hardware, generally offer orders of magnitude better energy efficiency than a general-purpose core [11].

Accelerator-rich architectures achieve peak performance and power efficiency by dynamically powering up only a small number of cores that best match the processing requirements of the given workload, with all other cores remaining dark when not in use. In this scheme, the set of active cores acts as a customized application-specific chip for the workload. In addition to better performance, the power efficiency of specialized cores allows powering up more cores, effectively pushing the power wall further than is possible with general-purpose cores [10].

The primary source of energy savings in ASIC accelerators is the elimination of the software overhead. When an application runs on a conventional processor, the software overhead includes instruction cache access, instruction fetch, instruction decode, instruction scheduling and reordering, and register file access. In a previous study, it has been shown that these steps account for more than 90% of the total energy consumption of a processor [16]. The other contributors to energy usage are the datapaths, i.e., data cache and functional units. A hardware accelerator eliminates software-related steps and only keeps the steps related to datapaths, i.e., data cache and datapath access. Datapath energy consumption of an accelerator is also lower than the datapaths inside a general-purpose processor, as the datapath is now customized for a specific application and has no extra logic (multiplexers, decoders, etc.) embedded in a typical functional unit for flexibility.

Accelerators are already becoming commonplace in current commercial mobile, desktop, and server processors and this trend is expected to continue in future [38]. The most common specialized cores include video/image/voice codecs, encryption engines, and standard interface controllers (Wi-Fi, USB, etc.), to name a few. With the emergence of the dark silicon, as mentioned earlier, current accelerator-enabled processors will transform to accelerator-dominated (or accelerator-rich) processors [38].

However, designing scalable accelerator-dominated processors will raise numerous architectural challenges. One major challenge is selecting the right set of accelerators for a many-core processor [16,40]. GreenDroid is an early attempt to synthesize specialized cores for key portions of the Android operating system [16]. Specialized cores are automatically generated from the C code of the widely used Android applications and can attain up to $10 \times$ more energy efficiency than a general-purpose core.

Nonetheless, designing ASICs for quickly changing general-purpose applications, particularly when applications are unknown at design time, is challenging and largely infeasible. To avoid the sophisticated core specialization procedure and get greater flexibility, some prior works design many-core processors with heterogeneous general-purpose cores [39]. In this approach, each core is general purpose, i.e., is capable of executing every application, but cores have different microarchitectures and processing powers. By choosing a right selection of the cores, programs with different degrees of parallelism can be executed efficiently [39]. To benefit from both flexibility and power efficiency, a many-core chip can come with embedded FPG-like fabric rather than dedicated fixed-function cores to take advantage of both run-time ASIC-like core specialization and flexibility [61]. These systems, however, suffer from the run-time FPGA reconfiguration complexity and cost.

Neural acceleration is a recent trend that utilizes artificial neural networks as accelerator [62–64] to provide an intermediate point between the efficiency of ASIC accelerators and the flexibility and generality of programmable cores.

Inspired by the human brain, an artificial neural network is an information processing model that is composed of a set of highly interconnected trainable processing elements (neurons). In the most common class of artificial neural networks, neurons are arranged in several layers. Input data is fed to the first layer (input layer) and processed as moves forward to the intermediate (hidden) layers and then to the last (output) layer, where the final output of the neural network is generated. Each neuron receives multiple inputs from its previous layer, calculates the weighted sum of its inputs, and sends the result to the neurons of the next layer. Each neuron has a set of weights, each corresponding to an input, which is derived from an offline learning phase. Unlike conventional computers that use an algorithmic approach to solve a problem, neural networks learn to solve a problem by an example, through a so-called learning process. The learning adjusts the weights of neurons for a specific application.

Neural networks were originally designed to solve classification, pattern recognition, time series prediction, and curve-fitting problems. However, it has been shown that neural networks can mimic and approximate any general function with acceptable accuracy. In particular, neural networks can accelerate compute-intensive functions, as they can replace the original code with a set of highly parallel multiply-and-accumulate operations. This kind of acceleration is best applied to functions that have application-level tolerance against inaccurate output, as neural network approximates (and does not perfectly imitate) the original code [62].

Neural Processing Unit (NPU) is the first attempt to use neural networks for acceleration [62]. NPU relies on an automated algorithmic transformation that converts compute-intensive segments of a code to neural network. This involves finding candidate code segments, generating neural network training data (by feeding many diverse input values to the code segment and recording the output), and training the neural network by the collected data to accelerate the code segment for which it is trained. Once the neural network is trained, the processor no longer executes the original code and, instead, executes the neural network trained for the code. As the authors of Ref. [62] report, many applications from different domains can take advantage of the neural acceleration of NPU.

Apart from core specialization, assembling many cores into a single many-core chip increases the complexity of intercore communication and memory access, so requires a scalable and efficient on-chip interconnect. Packet-based NoCs are envisioned to be the most efficient and scalable solution for connecting tens to hundreds of components in a future many-core processor. The next section discusses the challenges of such large NoCs and shows how the NoC can be specialized for a given set of active cores.

4. SPECIALIZED NoC FOR SPECIALIZED CORES

In a many-core processor in the dark silicon era, with a vast majority of on-chip cores left powered off (or dark), it is very likely that active cores should be noncontiguous and distributed across the network in an irregular way.

If the NoC is designed and customized for a single application (or a group of similar applications), the cores that best match the processing requirements of each task of the application are selected (or generated) first and then a mapping algorithm performs the core to NoC node mapping selectively.

The major objective of most mapping algorithms is to place the cores that communicate more often (and with high volume) close to each other [65]. However, modern general-purpose many-core processors employ a larger number of increasingly diverse applications (often unknown at design time) with potentially different traffic patterns. In such systems, it is not possible to find proper core-to-NoC mapping for all applications. For example, Fig. 1 shows the cores of a heterogeneous MPSoC that are activated to run two different applications. Each application activates and runs on those cores that best match its tasks. If the NoC was designed and customized for each individual application, the cores would be mapped into adjacent nodes. However, in a general-purpose platform, it is very likely that the preferred cores should be nonadjacent. The problem in this case is that the application and the intertask traffic may not be known at design time, when the mapping is performed. Even for the target applications that are specified a priori, it is often infeasible to find a mapping that is suitable for all applications, as intercore traffic pattern varies significantly across applications.

In this case, specialized cores for a particular application domain or class of applications (e.g., multimedia applications) are placed in a contiguous region of the chip area. However, the mapping of cores inside the region may not be optimal for all of the applications in that domain, as each application uses the cores with its own traffic pattern.

Fig. 2 shows a region of a many-core chip that is assigned to the cores required by a multimedia application set [20]. The application set contains H263 encoder and decoder and MP3 decoder and encoder. The mapping of cores inside the region is done by the algorithm we presented in a previous work [20] which is designed to map multiple applications with different intercore communication patterns. As Fig. 2 shows for two different applications of this domain, they activate different cores of the region and we still need topology optimization for such region-based NoCs.

Our previous study demonstrates that for two applications, namely x and y, that use the same set of cores but with 50% difference in intercore traffic, running x on an NoC whose mapping is optimized for y increases the communication latency by 30%–55%, compared to customizing mapping for x [10].

In a partially active CMP, conventional NoCs still necessitate all packets generated by active cores to go through the router pipeline in all intermediate nodes (both active and inactive) in a hop-by-hop basis. In this case, many packets may suffer from long latencies, if active nodes are located at a far topological distance.

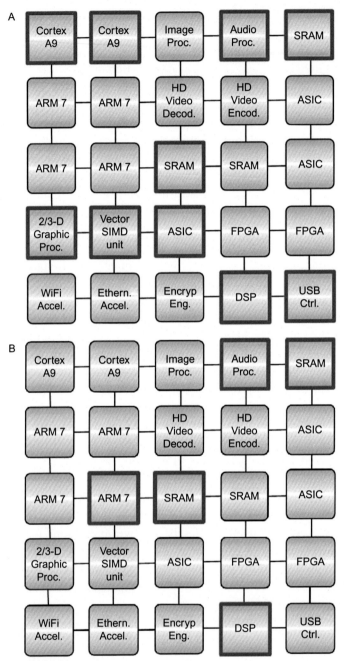

Fig. 1 A heterogeneous CMP with specialized cores. Active cores when playing game (A) and listening to music (B).

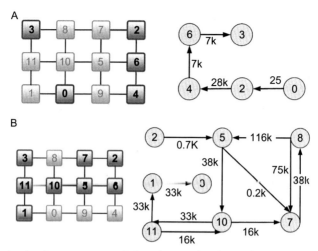

Fig. 2 A region in a heterogeneous CMP with specialized cores for the MMS applications and active cores of the region when running MP3 decoder (A) and H263 encoder (B) [10]. The communication task graph related to each application is also illustrated.

Due to the dynamic nature of core utilization pattern in such processors, in which the set of active cores varies over time, a reconfigurable topology is an appropriate option for adapting to the changes in network traffic. Among the existing reconfigurable topologies [20–23], we focus on the architecture we proposed in a previous work [20] (RecNoC, hereinafter), as it provides a more appropriate trade-off between flexibility and area overhead than the others. As mentioned before, RecNoC relies on embedding configuration switches into a regular NoC to dynamically change the interrouter connectivity. In this work, we show how the routers of the dark regions of a chip can be used as configuration switches and achieve the same level of power reduction and performance improvement as RecNoC, largely without paying its area overhead. The proposed NoC provides reconfigurable multihop intercore links among active cores by using the router of dark cores as bypass paths. This enables NoC to operate in the same way as a customized NoC, in which the cores are placed at nearby nodes.

Obviously, with the proposed reconfiguration, the physical distance of the cores is not changed. However, this topology adaption mechanism reduces communication power and latency significantly by reducing the number of intermediate routers. The latency and power usage of routers often dominate the total NoC latency and power usage. For example, in Intel's 80-core TeraFlops, routers account for more than 80% of the NoC power consumption, whereas links consume the remaining 20% [66].

Existing NoC topologies range from regular tiled-based [5,66,67] to fully customized structures [68–70]. Regular NoC architectures provide standard structured interconnects that often feature high reusability and short design time/effort. On the other hand, being tailored to the traffic characteristics of one or several target applications, customized topologies offer lower latency and power consumption when running that applications. However, topology customization will transform the regular structure of standard topologies into a nonreusable ad hoc structure with many implementation issues such as uneven wire lengths, heterogeneous routers, and long design time. Since our proposal realizes application-specific customized topologies on top of structured and regular components, it stands between these two extreme points of the topology design to get the best of both worlds; it is designed and fabricated as a regular NoC, but can be dynamically configured to a topology that best matches the core activation pattern of a partially active multicore processor.

In the next section, we review some related research proposals on NoC power/performance optimization and then present our dark-silicon aware NoC architecture.

5. LOW-LATENCY AND POWER-EFFICIENT NoC ARCHITECTURES

NoC power and latency reduction have been addressed in various proposals in the past mainly through reducing the hop count, per-hop latency, and blocking latency of network or by using low-latency power-efficient circuit-level technologies.

5.1 Network Hop Count Reduction

Packet hop count reduction has been the major way to reduce power and latency in many NoC designs. The reason is that when the average number of intermediate routers in a path reduces, the power consumption and latency overhead of routers reduce proportionally. Outstanding prior works that target hop count reduction include low-diameter high-radix topologies [67], reconfigurable networks [20–22], and ad hoc and semiregular topologies. Semiregular topologies are generated by adding extra links to a standard topology [71], whereas ad hoc topologies are generated from scratch either randomly [72] or based on application's traffic pattern [73]. As an example of a reconfigurable topology, in Ref. [21], a polymorphic NoC is presented that is constructed by selecting required components from a pool of configurable buffers, crossbars, and links. The network can be dynamically configured to offer the same performance as a fixed network, but with 40% area overhead.

Core to network mapping [69] is also an application-specific method that effectively reduces average hop count for a target application, when the application and its traffic pattern can be precharacterized at design time. These methods reduce the average packet hop count by placing the cores with high communication demand close to each other.

Few previous studies address the issue of topology optimization in the dark silicon era [74–76]. In Ref. [74], Cong and Xiao try to optimize the on-chip interconnect between cores and shared cache banks in a many-core dark silicon chip. The architecture they considered uses a crossbar as the underlying interconnect and their method configures core port-to-memory bank connections in such a way that the power efficiency of data access is maximized.

5.2 Blocking Latency Reduction

The network latency is broken down into blocking latency and link/router traversal latency (zero load latency). Unsuccessful virtual channel and switch allocation due to conflict with other packets are the sources of blocking latency.

Adaptive routing is the primary approach to reduce blocking latency by directing packets to less congested paths [77]. Among existing adaptive routing methods, those methods that make routing decision based on both local and global congestion metrics [78,79] or are aware of the running application's traffic behavior [80] can better balance the traffic across the network and reduce network hot spots.

Arbitration algorithm also plays an important role in reducing and balancing blocking latency. For example, Aeriga NoC leverages the system-level slack of a packet, defined as the amount of time the packet can be delayed in the network with no effect on execution time, to prioritize packets with a lower slack value during arbitration [81]. A similar application-aware arbitration is presented in Ref. [82] that ranks applications based on a stall-time criticality metric and has all routers in the network prioritize packets based on the applications' stall-time criticality.

The QoS level of packets [83] and message classes (request, response, coherency) [84] has also been used as the arbitration criteria in some prior work. In Ref. [85], it has been shown that by chaining packets destined to the same output port, the performance of arbitration increases.

Several works focus on reducing the probability of VC and switch arbitration failure by run-time adaptive buffer sizing [86] and heterogeneous resource allocation [87,88] in favor of network paths with high traffic. Switch arbitration failure rate can also be reduced by parallel short packet forwarding, which allows two short control packets to be forwarded in parallel on the same link [89], and bidirectional NoC, in which each regular NoC link can be dynamically reconfigured to transmit flits in either direction. In Ref. [90], unsuccessful switch allocation rate is effectively reduced by borrowing bandwidth from adjacent links for one cycle in the third dimension in a 3D NoC.

5.3 Per-Hop Latency Reduction

Per-hop latency is the time required to traverse the router pipeline stages and has a great impact on total NoC performance. To decrease this latency, prior methods often seek to cut down the pipeline stages of NoC routers. For example, look-ahead routing and speculative VC/switch allocation are two primary ways to reduce router pipeline stages [77]. Router bypassing is the primary approach to direct flits to preestablished paths inside routers and skip over the pipeline stages, effectively enabling one hop per cycle traversal. Several prior works, such as token flow control [91], pseudo-circuits [92], and virtual point-to-point connections [93], have shown that router bypassing can lead to considerable power/latency reduction.

In Ref. [94] it is observed that a flit can traverse 9 to 11 hops of a place-and-routed NoC design in 45 nm within a 1 GHz clock cycle. Based on this observation, the authors proposed SMART NoC (Single-cycle Multihop Asynchronous Repeated Traversal) [94]. SMART dynamically establishes single-cycle multihop bypass paths on regular NoC links and crossbars without adding any extra physical channels to bypass several intermediate routers. A multihop path is set up in a per cycle basis by a control network. In SMART, setting up bypass paths involves adding one cycle to the router pipeline stages. PRA eliminates this extra cycle by proactively setting up the multihop paths to further increase performance [95]. It leverages the packet blocking time (in which a packet is stalled in a router because the required resources are dedicated to another packet) and the time between the end of the tag and data lookup for data packets (that are sent in response to a request packet) for proactive path setup.

NoC-Out is an optimized interconnection for server applications and benefits from both the high-radix flattened butterfly topology to reduce hop count and low latency simple routers to reduce per-hop latency [96].

In NoC-Out, cores are grouped into clusters, interconnected by a backbone flattened butterfly topology. Each cluster is associated with a last-level cache slice, and a simple and fast ring-based NoC is used to connect the cores inside a cluster.

Bufferless NoC is a class of architectures that cut down router pipeline stages by always forwarding packets to some output port, even to a non-preferred one, in a single cycle [97,98]. This way, buffering, flow control, and complex arbitration are eliminated, so packets may be misrouted or dropped if their preferred output port is not available. This increases network latency under moderate and high traffic loads.

Circuit switching is an alternative NoC design and has been the main focus of many researches in the past [99,100]. Circuit-switched data traverse on reserved dedicated paths and need not go through buffering, routing, VC allocation, output arbitration, and flow control once a circuit is set up. However, circuit switching suffers from long circuit setup delay and poor bandwidth utilization.

The Time-Division Multiplexing scheme mitigates low bandwidth utilization of circuit switching [100] by dividing link bandwidth into multiple time slots and allocating each slot to a circuit. Although it increases bandwidth utilization, the high complexity of time slot allocation introduces new difficulties in using circuit switching.

Long setup time is also removed from the end-to-end packet latency by a proactive circuit setup. Déjà vu switching hides long circuit setup time by having request packets reserve a circuit for their anticipated response packet as they travel toward the destination [101]. Flit-reservation flow control [102] is another attempt to proactively reserve resources. To this end, a control flit traverses the network ahead of its corresponding data flit(s) to reserve buffers and channel bandwidth. This method uses a faster control network to guarantee that control flit always leads its corresponding data flits.

5.4 Low-Latency Power-Efficient Circuit Technologies

The emerging and unconventional circuit-level technologies have opened new opportunities for NoC power/latency reduction. Among them, optical [103,104], wireless [105], and 3D [106,107] networks-on-chip show more promise to replace the existing electrical interconnects.

Further, network-on-interposer is a new interconnect scheme that considerably improves the energy/performance profile of 2D chips by accelerating (and reducing the energy usage of) data transfer between cores and memory banks [57].

6. ARCHITECTURE SUPPORT FOR TOPOLOGY RECONFIGURATION

In this section, we first introduce the baseline RecNoC architecture and then show how the same reconfigurable topology can be implemented by the dark routers of a many-core chip.

6.1 RecNoC Architecture

RecNoC relies on simple configuration switches to dynamically change the internode connections of the NoC. In this architecture, routers (squares in Fig. 3) are not connected directly to each other, but through the configuration switches (circles in Fig. 3).

A configuration switch consists of some simple programmable internal connections between every incoming and outgoing links of the switch. A customized topology is generated by appropriately setting the internal connections of configuration switches to make a new set of links between NoC routers. The topology is actually generated by chaining some regular links via these switches to construct longer links with different lengths. To prevent reconfigurable long links, established by chaining multiple regular links, from degrading the NoC clock frequency, some switches segment the reconfigurable long links into fixed length links by buffering incoming flits and forwarding them in the next cycle in a pipelined fashion. The details about optimizing the switch structure and important implementation issues can be found in Ref. [20].

This NoC can be configured to implement arbitrary and standard topologies by appropriately setting the configuration switch connections. RecNoC adapts the topology to the on-chip traffic pattern by calculating

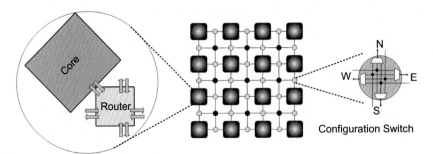

Fig. 3 The RecNoC architecture [20].

the proper topology for each application offline and loading it into the network when the corresponding application starts. Alternatively, it can calculate the proper topology at run-time using an online light-weight traffic monitoring and configuration calculation process. The algorithm for mapping a multitask application onto RecNoC and customizing the topology for it is available elsewhere [20,24].

The reconfigurable topology depicted in Fig. 3 can be extended to a more general and customizable architecture shown in Fig. 4. This architecture offers a hierarchical clustered communication infrastructure with two parameters: cluster size, i.e., the number of nodes of each cluster, and corridor width, i.e., the number of configuration switches between adjacent clusters. This architecture can be considered as a hierarchical topology in which several mesh subnetworks are connected via a higher layer topology implemented on reconfigurable switches. The reconfiguration capability is the key advantage of our proposal over the existing hierarchical NoC architectures.

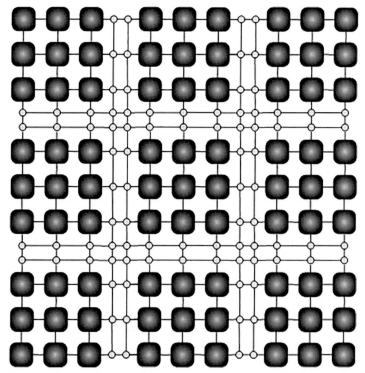

Fig. 4 A cluster-based architecture for reconfigurable NoCs with the cluster size of 9 and corridor width of 2.

In Fig. 4, the NoC nodes are grouped into 9-node clusters and the corridor width is 2. The connection among the nodes within a cluster is fixed and the clusters are interconnected via configuration switches. The idea behind this structure is to benefit from the interesting characteristics of the mesh topology, while avoiding its long diameter. From the traffic management perspective, a mesh NoC is efficient in handling local traffic patterns where each node communicates more with the nodes in local neighborhood. This traffic pattern commonly occurs when a mapping algorithm is applied, as most mapping methods aim to map the frequently communicating tasks near each other. Nonetheless, when the network size grows, some packets may suffer from long latencies of mesh due to the lack of direct paths between remotely located nodes.

The proposed architecture can efficiently support local traffic by mapping the tasks with high communication demand into the same cluster. In this cluster-based design, the local traffic will not pass through configuration switches to further save energy. The long diameter of large mesh networks can be also mitigated by configuring the intercluster connections in such a way that a direct connection (path with reduced hop count) is constructed between the endpoint nodes of long-distance traffic flows.

In Fig. 4, the nodes of a cluster are interconnected by a conventional mesh. This intracluster topology, however, can be replaced with several other topologies. In particular, when the cluster size is small, a shared bus can efficiently handle the internal traffic of clusters.

6.2 Dark Silicon-Aware Reconfigurable NoC

In this section, we propose to leverage unused dark routers as configuration switches to get a reconfigurable NoC with lower cost. To this end, we slightly modify the architecture of a conventional NoC router [77], in order to add the capabilities of a configuration switch to it. The modified routers are then used to customize the network connectivity among the active cores.

6.2.1 Baseline Wormhole-Switched Router

A conventional wormhole-switched router forwards packets in a pipelined fashion [77]. The first stage of this pipeline is the buffer write (BW) stage, in which the header flit, on arriving at an input port, is buffered in an input buffer. In the next stage, the route computation (RC) stage, the routing logic makes a routing decision to assign an output port to the header flit (and its packet). The header flit then requests a virtual channel from the virtual-channel allocator of the selected output port in the virtual-channel allocation (VCA) stage. The RC and VCA stages are done in a per-packet basis, but the remaining stages are done for each individual flit. The next stage is switch allocation (SA), in which the flit requests for a single-flit time slot on a crossbar to move to the output selected at the RC stage. In case there are multiple active virtual channels, this allocation may involve competition not only for the switch output port, but also with other virtual channels in the same input port for the switch input. If the switch allocation is successful, the flit completes switch traversal in the next stage (ST), followed by the link traversal (LT) stage to reach the downstream router. Body flits follow the same pipeline stages except that they do not go through RC and VA stages. Finally, when the tail flit of a packet leaves the router, it frees the resources reserved by the header flit.

6.2.2 The New Router Design

Fig. 5 shows the architecture of the modified router. Every input port has an internal shortcut path that allows packets skip the buffering, routing, switch, and VC allocation pipeline stages and directly head for the crossbar. A register (REG in Fig. 5) is added to the shortcut paths to buffer incoming flits at each hop for one cycle, effectively allowing reconfigurable long link

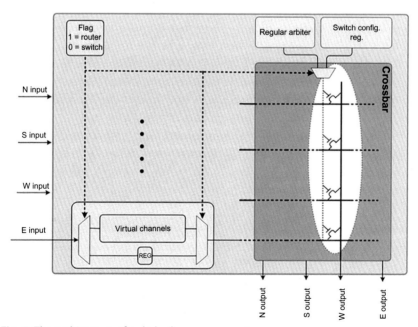

Fig. 5 The architecture of a dark silicon-aware router.

traversal in a pipelined manner. We conservatively buffer flits at each hop along a reconfigurable long link to guarantee that NoC timing constraints are not violated. Nonetheless, prior works observed that based on the NoC clock frequency and datapath delays, flits may pass several hops in a single cycle without the need for buffering [94]. By these shortcut paths, a reconfigurable long link is formed between two remote active nodes.

The new design also involves modifying the switch allocator; it is now provided with an extra register per each output port that keeps the fixed crossbar input–output connections when the router acts as a configuration switch. If the router is functioning normally, the switch output is allocated by the regular wormhole-switching arbiter. Otherwise, if the router is downgraded to a configuration switch, this register determines whether the corresponding output port is part of a reconfigurable long link and if so, specifies which input port should be directly connected to it. This register is set when the topology is constructed and remains fixed until the next topology reconfiguration event.

In a conventional router, as mentioned earlier, flits are first buffered at each hop and then go through the router pipeline stages. However, when the routers are bypassed, flits are directly forwarded to a predefined output through the fixed crossbar connections. Consequently, they are not required to go through the buffering and flow control, routing, VC allocation, and switch arbitration pipeline stages and just pass through the crossbars and links, which cover the physical distance between the end point nodes.

We modeled the proposed router in VHDL and synthesized the code using a commercial synthesis and physical design tool in 45 nm technology. The result shows a negligible increase in the area of the modified router (less than 5%) compared to a conventional router.

In Fig. 6, we show how the reconfigurable long links are configured to construct a mesh topology for a set of active cores. In addition to standard topologies, the proposed architecture is capable to construct irregular customized topologies. Fig. 7B and C shows how the reconfigurable RecNoC, depicted in Fig. 3, and the NoC proposed in this chapter, respectively, can be customized for an application whose communication task graph is depicted in Fig. 7A.

Assuming an accelerator-rich CMP, where each task is allowed to (or gives the best performance when) run on a specific core, the topology is reconfigured in such a way that the frequently communicating nodes have the minimum topological distance. Ideally, all communicating cores will be provided by a direct reconfigurable long link. In practice, however, we often

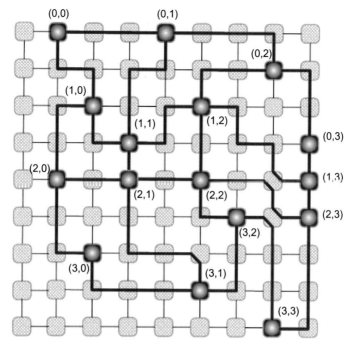

Fig. 6 Constructing a mesh topology for the active cores of an accelerator-rich CMP. Each active core is colored in *red* and marked by its mesh address.

cannot construct such ideal topology: the application may have many communication flows (if a large portion of cores are active) and the limited inactive routers cannot afford to provide direct connection for all communicating pairs. In this case, some communication flows are provided by multihop paths. For example, the data transferred between nodes 1 and 5 in Fig. 7A is routed through node 6 in the topology generated in Fig. 7B. Note that the proposed reconfigurable NoC still routes packets as a wormhole-switched network, but its topology is customized for the on-chip traffic.

7. TOPOLOGY RECONFIGURATION PROCEDURE

In this section, we present an algorithm to establish a customized topology using a given set of active cores for an input application.

The input applications are modeled by a *communication task graph* (CTG). CTG is a directed graph $G(V,E)$, where each $v_i \in V$ represents a task, and a directed edge $e_{i,j} \in E$ represents the communication flow from v_i to v_j with

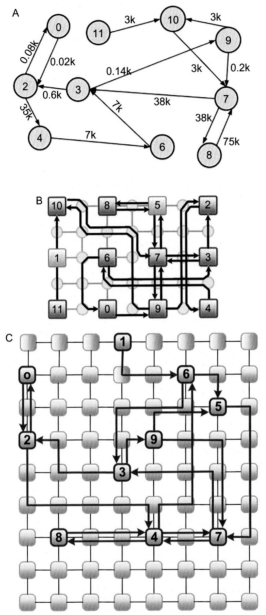

Fig. 7 Optimizing the topology for the task graph of a multimedia application (A) on RecNoC (B) and the proposed NoC (C). The *red cores* in the proposed NoC represent active cores.

the communication volume of $t(e_{i,j})$. A sample CTG is illustrated in Fig. 7A. The subset of NoC cores (and associated routers) that are activated to execute the tasks of an incoming application is represented by *ActiveC*, while the remaining cores comprise the *InactiveC* set. The routers of inactive cores are used as configuration switches. When acting as a switch, the main router logic of them is turned off to save static energy and just the internal shortcut paths are active.

The algorithm takes the CTG and *ActiveC* of an input application as input and constructs a virtual topology on top of the physical mesh topology through reconfigurable long links by leveraging the routers of *InactiveC* such that:

$$\text{Min}\left\{\sum_{\forall e \in E} t(e_{i,j}) \times Hop(v_i, v_j)\right\}$$

where $Hop(v_i,v_j)$ is the hop count (the number intermediate routers) between nodes v_i and v_j in the new topology.

Topology selection can be carried out either offline or online. In the offline mode, the algorithm finds an appropriate topology a priori and stores it with the application. The topology then will be loaded onto the network once the application starts. The online scheme, on the other hand, is especially beneficial when multiple applications are running simultaneously and new applications can be inserted incrementally. In this work, we focus on the offline procedure.

7.1 Topology Reconfiguration Algorithm

Fig. 8 outlines the proposed topology reconfiguration algorithm. Initially, all routers of *InactiveC* are at the unconfigured state with all internal connections open. Algorithm starts by selecting the CTG edges in the decreasing order of communication volume (*FindMaxEdge()* function in Fig. 8), and then a modified version of the Dijkstra's shortest path algorithm is carried out to find a path with minimum weight between the endpoint nodes of the edge (*Dijkstra()* function). In this procedure, the network (including both *InactiveC* and *ActiveC* nodes) is considered as a directed weighted graph and the cost of a path is calculated as the cumulative cost of the routers (*ActiveC* routers) and configuration switches (*InactiveC* routers) that are chained to construct the path. We assign a cost of 1 to a link ending to a configuration switch and a cost of 3 to a link ending to a router. These costs reflect the power/latency ratio of a configuration switch and a router obtained through synthesizing the VHDL description of these components.

Algorithm 1. Topology reconfiguration algorithm

App: Input application

$G(V,E)$= Communication task graph (CTG) of *App*

N: The NoC with *m×n* mesh topology

InactiveC : The set of active nodes of *N* that are turned on by the tasks of *App*

ActiveC: The set of inactive nodes of *N* that are not needed during *App*'s execution

for all e ∈ E **loop**// *the main loop*
    E_{max}= FindMaxEdge(G); //*select the CTG edge with maximum bandwidth*
                          demand
    P = Dijkstra(N, E_{max}.Source, E_{max}.Dest);
    If (P = NULL)// *No path is found*
        Conflict_vect=Dijkstra_Min_Conflict(N, E_{max}.Source, E_{max}.Dest);
        for (all elements of Conflict_vect) **loop**
            revert Conflict_vect(i++) to a router;
            P = Dijkstra(N, E_{max}.Source, E_{max}.Dest);
            If (P)
                Break;
            end if;
        end for;
    end if;
    for all nodes ∈ *InactiveC* **loop**
        Configure the shortcut path inside the router to setup P
    end for
    for all nodes ∈ *InactiveC* **loop**
        Set the routing table of the router to setup P
    end for
end for// *the main loop*

Fig. 8 Topology reconfiguration algorithm.

Assigning higher cost to routers encourages the algorithm to find a path through the configuration switches, hence facilitating in creating longer reconfigurable links for communication flows.

In this algorithm, reconfigurable long links are not allowed to conflict; so, if a configuration switch is already configured in previous iterations of the algorithm for the flows with higher traffic rates (e.g., by connecting its N input port to S output port), the algorithm is not allowed to use turns that conflict with already configured inputs (e.g., connecting N input port to W output port) and outputs (e.g., connecting E input port to S output port) on that switch.

To restrict Dijkstra's algorithm to find topological shortest paths (and not any path with minimum cost), it is only allowed to use the links that progress toward the destination node, i.e., the links that reduce the distance to the destination by one hop at each step. Further, the algorithm is not allowed to violate the maximum bandwidth of NoC links to prevent congestion on links.

If the current configuration of the configuration switches blocks all paths between the endpoint nodes of a not yet mapped CTG edge, the algorithm resolves the conflict in three steps: (1) selects the unsuccessful path found for the CTG edge that has the minimum number of conflicting configuration switches (switches whose configuration prevent the path advancing toward the destination), (2) selects one of the conflicting switches along that path and makes it revert to a router, and then (3) retries path finding by Dijkstra's algorithm.

The algorithm examines conflicting switches (step 2) in the path found in step 1 one by one until a valid path is found. If the problem persists, the algorithm returns to step 1 and selects the next minimum conflicting path. When a configuration switch reverts to a router, it can make connections between any two ports (as it has buffering and arbitration) and has no restrictions associated with switches.

This three-step procedure guarantees finding always a path, because at the worst case, all configuration switches revert to routers and the problem, consequently, converts to finding a path in a regular mesh.

In this complementary stage, the procedure repeats the Dijkstra's algorithm (*Dijkstra_Min_Conflict*) for step 1 to find the unsuccessful path with minimum cost, but this time, the cost of using a configuration switch against its current configuration is not infinity, but a large integer. As a result, the algorithm will always return a path in step 1. This path has the minimum number of conflicting configuration switches. The algorithm also returns the conflicting configuration switches in a vector (*Conflict_vect*), where the conflicting switches are listed in the order at which they appear in the path. This vector is used in the second step of this procedure to identify conflicting switches that should revert to a router.

When a configuration switch reverts to a router, all reconfigurable long links that already pass through it (found in previous rounds of the algorithm) are broken into two parts. To keep the connection, the routing table of the router is set appropriately to bridge the two reconfigurable link parts via the router.

After a path is found for a CTG edge, all configuration switches and routing tables of the path are configured accordingly and the algorithm continues with the next flow. Configuration switch is configured by appropriately setting the select lines of the multiplexer of the input units and the output allocator registers to fix the crossbar input–output connections.

A path may contain several routers and reconfigurable long link segments. For example, the path between nodes 7 and 8 in Fig. 7B consists of two reconfigurable long links (7 → 4 and 4 → 8) and an intermediate router. Packets traverse the first reconfigurable long link between nodes 7 and 4, enter the router and pass through the pipeline stage in node 4, and then take the next reconfigurable long link to bypass the routers between nodes 4 and 8 to reach the destination.

7.2 Topology Reconfiguration for CMP Workloads

The proposed architecture and algorithm can support both CMP and multicore SoC workloads. Multicore SoCs are generally embedded special-purpose processors. In such systems, the applications and their traffic characteristics are often known at design time. Further, each input application can often be spatially partitioned into several parallel tasks, each of which is assigned to a specialized processing unit. The communication pattern among cores remains relatively static as each core executes a fixed code. The CTG of each application provides the communication pattern (source–destination pairs) and estimates of bandwidth demands between communicating pairs and can be obtained through either program profiling or code analysis.

On the other hand, most CMP applications are composed of multiple similar tasks that all execute the same code, but on different data. As these applications are often designed based on the shared-memory model, the source of on-chip traffic in such applications is data block transfers initiated by read/write operations on the shared data. Therefore, all source-destination pairs can potentially exchange packets. This necessitates the existence of at least one path between any source–destination pair of the network when running these workloads.

Although the proposed algorithm works well for sparse CTGs, its execution time increases considerably for fully connected CTG of typical CMP workloads. As mentioned earlier, we should guarantee the existence of at least one path between all active source–destination pairs when running CMP workloads with fully connected or dense CTGs. Further, these applications often distribute the traffic load evenly across the NoC nodes, since they balance the processing load across all cores, so the bandwidth demand between all source–destination pairs is rather the same. To efficiently handle such fully connected CTGs, we slightly modify the baseline algorithm described earlier by allowing the CTG edges to reuse already established reconfigurable long links. To this end, when a long link is established for a CTG edge between two active cores x and y, the source core (x) checks the already established connections of y to see if it can reuse them for its edges that are not yet selected for long link construction.

Reusing already established long links to decrease the problem size for fully connected or dense CTGs is implemented as an extension to the baseline algorithm outlined in Fig. 8. Assume that core A in Fig. 9 runs a task that has seven connections to cores B, C, D, E, F, G, and H, and among them, the A ➜ B connection (CTG edge) has the biggest bandwidth demand. The algorithm will consider the A ➜ B connection at some iteration and establish a reconfigurable long link for it (shown by the dashed line in Fig. 9). At this point, the extension of the algorithm starts. It compares all nodes that are connected directly to B by a reconfigurable long link (nodes C, D, and E in Fig. 9) against the outgoing CTG edges of node A that have not yet been mapped by the algorithm (i.e., connections to C, D, E, F, G, and H). Then, the matched nodes (nodes C, D, and E) are selected one by one, and for each node, for example, node C, the algorithm checks whether A ➜ B ➜ C path is a shortest path between nodes A and C. This is checked by calculating the Manhattan distance between nodes A and C and comparing it to the A ➜ B ➜ C path length. If it is a shortest path, the path will be registered as the solution for the A ➜ C connection.

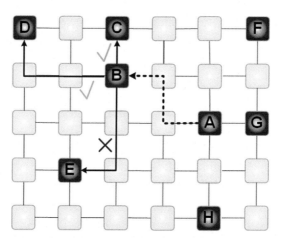

Fig. 9 Node A, upon constructing the A ➜ B reconfigurable long link can reuse the already established B ➜ D and B ➜ C reconfigurable long links for A ➜ D and A ➜ C connections.

In Fig. 9, node A can use the connections of node B to nodes C and D to forward packets destined to C and D through B. Thus, it can remove A ➔ C and A ➔ D from the list of the CTG edges waiting for connection construction and greatly reduce the problem size. However, it cannot reuse the B ➔ E connection, as the A ➔ B ➔ E connection is not a shortest path from A to E. Our results show that by using this long link reuse, the algorithm execution time reduces by about 60%, on average.

8. EVALUATION

We evaluate the proposed NoC architecture using several synthetic and realistic traffics of typical CMPs and multicore SoCs. Simulations are done using BookSim 2.0, a cycle-accurate simulator that accurately models a wormhole pipelined router [108]. Power consumption is calculated by the DSENT power model [109]. The routers are 64-bit wide with 8-flit buffers and two virtual channels per port. For CMP benchmarks, the simulator input is the traffic traces extracted during the program execution on a full-system simulator. For SoC workloads, we generate the traffic of each benchmark based on its CTG. To this end, packets are generated with exponential distribution and the communication rate between nodes is proportional to the communication volume between their corresponding tasks in the CTG.

The proposed NoC architecture is compared to a conventional NoC. The conventional NoC implements a three-stage pipelined wormhole router with route computation, speculative switch and VC allocation [77], and link and crossbar traversal. For more detailed experimental results, we refer the reader to Ref. [24].

8.1 CMP Workloads

We evaluate the proposed NoC using Splash programs [110]. Each program has 49 parallel tasks that run on a 12×12 network. In these experiments, the tasks are randomly mapped onto 49 NoC nodes (and leave the rest powered off) and then a customized topology is constructed for active cores. Fig. 10 shows the average packet latency for the proposed reconfigurable NoC, the conventional NoC, and a conventional NoC with the same size as the application. All network sizes are 12×12. The energy consumption results are shown in Fig. 11.

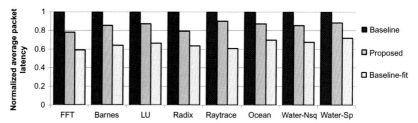

Fig. 10 Normalized average packet latency for the conventional NoC, the proposed reconfigurable NoC, and a conventional NoC that fit the application in size, under multicore SoC workloads.

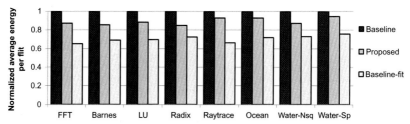

Fig. 11 Normalized average energy per flit for the conventional NoC, the proposed reconfigurable NoC, and a conventional NoC that fit the application in size, under multicore SoC workloads.

As the figure indicates, the proposed reconfigurable NoC outperforms the conventional NoC across all benchmarks and reduces packet latency by 15%, and average energy per flit by 11%, on average.

Fig. 11 also compares the energy and latency of the applications with the energy and latency that would be obtained on an ideal NoC that fits the application size (7×7). According to the results, the proposed reconfiguration can efficiently bridge the power/performance gap between a large accelerator-rich NoC in the dark silicon era and an NoC that fits the application size.

8.2 Multicore SoC Workloads

To further evaluate the performance of the proposed topology reconfiguration, we use some existing SoC programs as benchmark. The benchmarks are Multi-Window Display (MWD) with 12 tasks [111], Video Object Plane Decoder (VOPD) with 16 tasks [73], GSM decoder with 43 tasks [112], a real-time multimedia system (D26_media) with 26 tasks [113], and MMS (Multimedia System) with 40 tasks [71]. Again, we assume that the customized cores for different tasks of an application are distributed randomly throughout

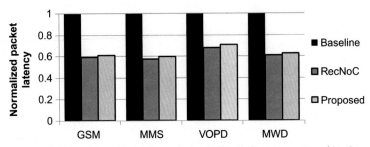

Fig. 12 Normalized average packet latency for RecNoC, the conventional NoC, and the proposed reconfigurable NoC under multicore SoC workloads.

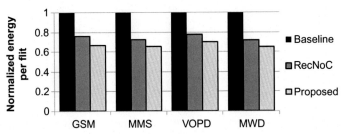

Fig. 13 Normalized average energy per flit for RecNoC, the conventional NoC, and the proposed reconfigurable NoC under multicore SoC workloads.

the network. The network size for all considered NoCs is 10×10 for GSM and MMS, 5×6 for MWD, 6×7 for VOPD, and 7×8 for D26_media. Here, we construct a customized topology for the active cores of each application based on its CTG.

Figs. 12 and 13 compare the three considered NoC architectures in terms of average packet latency and energy per flit, respectively. As the figures indicate, the proposed NoC significantly outperforms the conventional NoC and is comparable in performance with RecNoC which uses 21%–35% more hardware. As the number of communication flows in these benchmarks is by far less than the number of flows in CMP benchmarks, the algorithm can find better topologies for these benchmarks; hence more latency and energy reduction are achieved. The reduction in packet latency and energy consumption is 39% and 33%, respectively.

8.3 Sensitivity to Ratio of Active to Dark Cores

Obviously, the proposed method would offer more performance gains when the dark portion of the chip increases because there are more resources

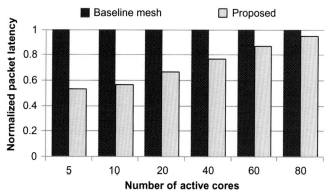

Fig. 14 Latency improvement of the proposed NoC over the conventional NoC for different number of active cores in a 10×10 NoC.

that can be exploited to optimize the topology. Fig. 14 studies the effect of the ratio of active to dark cores on the performance improvement of the proposed NoC in a 10×10 NoC under uniform traffic. The figure shows latency improvement of the proposed NoC over the conventional NoC for different number of active cores. The active cores are selected randomly. As the figure shows, even with 80 active cores, the proposed NoC shows 4% improvement in latency over the baseline. As it is predicted that more than half of the die area in future technologies will be dark, the proposed NoC can effectively increase network performance.

9. CONCLUSION

In this chapter, we presented a dark silicon-aware reconfigurable NoC. The design leverages the routers of the dark portion of the chip to customize the NoC topology for active cores. This mechanism primarily aims to improve the latency and energy consumption of the NoC. The new topology is obtained by bypassing the routers associated with dark cores to directly connect remote active routers of the topology. This customized topology can reduce the NoC average message hop count, which, in turn, results in a proportional reduction in the communication latency and power consumption of on-chip communication.

For a given application, we developed an algorithm to reconfigure the connectivity among active cores and obtained an application-specific irregular topology, optimized for the workload traffic characteristics. The experimental results, using both CMP and SoC workloads, showed that the proposed reconfiguration scheme improved network resource usage, energy consumption, and latency, while requiring only minor changes to the router and network architecture. It also makes more appropriate trade-off between flexibility and area overhead compared to RecNoC.

REFERENCES

[1] W.M. Holt, in: Moore's law: a path going forward, Proceedings of IEEE International Solid-State Circuits Conference (ISSCC), San Francisco, CA, 2016, pp. 8–13.
[2] C. Gonzalez, et al., in: POWER9™: a processor family optimized for cognitive computing with 25Gb/s accelerator links and 16Gb/s PCIe Gen4, Proceedings of IEEE International Solid-State Circuits Conference, ISSCC, San Francisco, CA, 2017, pp. 50–51.
[3] Intel® Xeon® Processor E7 Family, http://www.intel.com, 2017.
[4] K. Aingaran, et al., M7: Oracle's next-generation Sparc processor, IEEE Micro 35 (2) (2015) 36–45.
[5] A. Sodani, et al., Knights landing: second-generation Intel Xeon Phi product, IEEE Micro 36 (2) (2016) 34–46.
[6] S. Borkar, in: Thousand core chips: a technology perspective, Proceedings of Design Automation Conference (DAC), 2007, pp. 746–749.
[7] ISSCC 2017 Tech Trends, http://isscc.org/trends/, 2017.
[8] N. Hardavellas, M. Ferdman, B. Falsafi, A. Ailamaki, Toward dark silicon in servers, IEEE Micro 31 (4) (2011) 6–15.
[9] H. Mair, et al., in: A 10nm FinFET 2.8GHz tri-gear deca-core CPU complex with optimized power-delivery network for mobile SoC performance, Proceedings of IEEE International Solid-State Circuits Conference, ISSCC, San Francisco, CA, 2017, pp. 56–57.
[10] H. Esmaeilzadeh, et al., in: Dark silicon and the end of multicore scaling, Proceedings of ISCA, 2011, pp. 365–376.
[11] M.B. Taylor, A landscape of the new dark silicon design regime, IEEE Micro 33 (5) (2013) 8–19.
[12] M.B. Taylor, in: Is dark silicon useful?, Proceedings of Design Automation Conference (DAC), 2012.
[13] H.E. Sumbul, K. Vaidyanathan, Q. Zhu, F. Franchetti, L. Pileggi, in: A synthesis methodology for application-specific logic-in-memory designs, Proceedings of Design Automation Conference (DAC), 2015, pp. 1–6.
[14] R. Balasubramonian, et al., Near-data processing: insights from a MICRO-46 workshop, IEEE Micro 34 (4) (2014) 36–42.
[15] B. Falsafi, et al., Near-memory data services, IEEE Micro 36 (1) (2016) 6–13.
[16] N. Goulding-Hotta, et al., The GreenDroid mobile application processor: an architecture for silicon's dark future, IEEE Micro 31 (2) (2011) 86–95.
[17] J. Cong, M.A. Ghodrat, M. Gill, B. Grigorian, K. Gururaj, G. Reinman, in: Accelerator-rich architectures: opportunities and progresses, Proceedings of Design Automation Conference (DAC), 2014, pp. 1–6.
[18] P. Lotfi-Kamran, M. Modarressi, H. Sarbazi-Azad, in: Near-ideal networks-on-chip for servers, Proceedings of International Symposium on High Performance Computer Architecture, 2017.
[19] On-Chip Communications Network Report, Sonics Inc., 2013. www.sonics.com.
[20] M. Modarressi, et al., Application-aware topology reconfiguration for on-chip networks, IEEE Trans. Very Large Scale Integr. VLSI Syst. 19 (11) (2011) 2010–2022.

[21] M. Kim, J. Davis, M. Oskin, T. Austin, in: Polymorphic on-chip networks, Proceedings of ISCA, 2008, pp. 101–112.

[22] B. Zafar, J. Draper, T.M. Pinkston, in: Cubic ring networks: a polymorphic topology for network-on-chip, Proceedings of 39th International Conference on Parallel Processing, 2010, pp. 443–452.

[23] Z. Qian, S.M. Abbas, C.Y. Tsui, FSNoC: a flit-level speedup scheme for network on-chips using self-reconfigurable bidirectional channels, IEEE Trans. Very Large Scale Integr. VLSI Syst. 23 (9) (2015) 1854–1867.

[24] M. Modarressi, H. Sarbazi-Azad, Leveraging dark silicon to optimize networks-on-chip topology, J. Supercomput. 71 (9) (2015) 3549–3566.

[25] N. Hardavellas, The Rise and Fall of Dark Silicon, USENIX, 2012.

[26] L. Wang, K. Skadron, Implications of the power wall: dim cores and reconfigurable logic, IEEE Micro 33 (5) (2013) 40–48.

[27] A. Pedram, S. Richardson, M. Horowitz, S. Galal, S. Kvatinsky, Dark memory and accelerator-rich system optimization in the dark silicon era, IEEE Des. Test Comput. 34 (2) (2017) 39–50.

[28] J. Hennessey, D. Patterson, Computer Architecture: A Quantitative Approach, Morgan Kaufmann, 2011.

[29] J. Allred, S. Roy, K. Chakraborty, in: Designing for dark silicon: a methodological perspective on energy efficient systems, Proceedings of International Symposium on Low Power Electronics and Design, 2012, pp. 255–260.

[30] J. Henkel, H. Khdr, S. Pagani, M. Shafique, in: New trends in dark silicon, Proceedings of Design Automation Conference (DAC), 2015, pp. 1–6.

[31] G. Yan, Y. Li, Y. Han, X. Li, M. Guo, X. Liang, in: AgileRegulator: a hybrid voltage regulator scheme redeeming dark silicon for power efficiency in a multicore architecture, IEEE International Symposium on High-Performance Comp Architecture, 2012, pp. 1–12.

[32] T.S. Muthukaruppan, M. Pricopi, V. Venkataramani, T. Mitra, S. Vishin, in: Hierarchical power management for asymmetric multi-core in dark silicon era, Proceedings of Design Automation Conference (DAC), 2013, pp. 1–9.

[33] Qualcomm Snapdragon 810, https://www.qualcomm.com/products/snapdragon/processors/810.

[34] P. Hammarlund, et al., Haswell: the fourth-generation Intel core processor, IEEE Micro 34 (2) (2014) 6–20.

[35] AMD Phenom™ II Processors, http://www.amd.com/en-us/products/processors/desktop/phenom-ii, 2017.

[36] W. Wang, E.A. León, in: Evaluating DVFS and concurrency throttling on IBM's Power8 architecture, Proceedings of International Conference for High Performance Computing, Networking, Storage and Analysis, 2015.

[37] Cortex-A Series Programmer's Guide, http://infocenter.arm.com.

[38] Q. Zheng, N. Goulding-Hotta, S. Ricketts, S. Swanson, M. Bedford Taylor, J. Sampson, Exploring energy scalability in coprocessor-dominated architectures for dark silicon, ACM Trans. Embed. Comput. Syst. 13 (2014) 130.

[39] Y. Turakhia, B. Raghunathan, S. Garg, D. Marculescu, in: HaDeS: Architectural synthesis for heterogeneous dark silicon chip multi-processors, Proceedings of Design Automation Conference (DAC), 2013, pp. 1–7.

[40] Y. Zhang, L. Peng, F. Xin, H. Yue, in: Lighting the dark silicon by exploiting heterogeneity on future processors, Proceedings of Design Automation Conference (DAC), 2013, pp. 1–7.

[41] big.LITTLE technology, http://www.arm.com, 2017.

[42] NVIDIA Tegra 4 Family CPU Architecture 4-PLUS-1 Quad core, http://www.nvidia.com, 2017.

[43] K. Itoh, A historical review of low-power, low-voltage digital MOS circuits development, IEEE Solid-State Circuits Mag. 5 (1) (2013) 27–39.

[44] S. Chen, et al., Powering up dark silicon: mitigating the limitation of power delivery via dynamic pin switching, IEEE Trans. Emerg. Top. Comput. 3 (4) (2015) 489–501.

[45] S. Chen, S. Irving, L. Peng, H. Yue, Y. Zhang, A. Srivastava, Using switchable pins to increase off-chip bandwidth in chip-multiprocessors, IEEE Trans. Parallel Distrib. Syst. 28 (2017) 274–289.

[46] ITRS, International Technology Roadmap for Semiconductors, 2010 Update, http://www.itrs.net, 2011.

[47] A.N. Udipi, N. Muralimanohar, N. Chatterjee, R. Balasubramonian, A. Davis, N.P. Jouppi, in: Rethinking DRAM design and organization for energy-constrained multi-cores, Proceedings of ISCA, 2010.

[48] D. Zhang, N. Jayasena, A. Lyashevsky, J.L. Greathouse, L. Xu, M. Ignatowski, in: TOP-PIM: throughput-oriented programmable processing in memory, Proceedings of International Symposium on High-Performance Parallel and Distributed Computing, 2013, pp. 85–98.

[49] P. Emma, et al., in: 3D stacking of high-performance processors, Proceedings of HPCA, 2014.

[50] D.H. Woo, et al., in: An optimized 3D-stacked memory architecture by exploiting excessive, high-density TSV bandwidth, Proceedings of HPCA, 2010.

[51] L.W. Schaper, S.L. Burkett, S. Spiesshoefer, G.V. Vangara, Z. Rahman, S. Polamreddy, Architectural implications and process development of 3-D VLSI Z-axis interconnects using through silicon vias, IEEE Trans. Adv. Packag. 28 (3) (2005) 356–366.

[52] H. Tajik, et al., in: VAWOM: temperature and process variation aware wear out management in 3D multicore architecture, Proceedings of DAC, 2013.

[53] G.H. Loh, et al., in: A processing in memory taxonomy and a case for studying fixed-function pim, Proceedings of Workshop on Near-Data Processing (WoNDP), 2013.

[54] J. Ahn, S. Yoo, O. Mutlu, K. Choi, in: PIM-enabled instructions: a low-overhead, locality-aware processing-in-memory architecture, Proceedings of ISCA, 2015, pp. 336–348.

[55] J. Ahn, S. Hong, S. Yoo, O. Mutlu, K. Choi, in: A scalable processing-in-memory accelerator for parallel graph processing, Proceedings of ISCA, 2015, pp. 105–117.

[56] D. Kim, J. Kung, S. Chai, S. Yalamanchili, S. Mukhopadhyay, in: Neurocube: a programmable digital neuromorphic architecture with high-density 3D memory, Proceedings of ISCA, 2016, pp. 380–392.

[57] N.E. Jerger, A. Kannan, Z. Li, G.H. Loh, in: NoC architectures for silicon interposer systems: why pay for more wires when you can get them (from your interposer) for free?, Proceedings of International Symposium on Microarchitecture, Cambridge, 2014, pp. 458–470.

[58] D.U. Lee, et al., in: A 1.2 V 8Gb 8-channel 128GB/s high-bandwidth memory (HBM) stacked DRAM with effective microbump I/O test methods using 29nm process and TSV, Proceedings of IEEE International Solid-State Circuits Conference (ISSCC), 2014.

[59] Hybrid Memory Cube Specification 2.0, http://www.micron.com/products/hybrid-memory-cube, 2015.

[60] A. Boroumand, S. Ghose, B. Lucia, K. Hsieh, K. Malladi, H. Zheng, O. Mutlu, LazyPIM: an efficient cache coherence mechanism for processing-in-memory, IEEE Comput. Archit. Lett. 16 (1) (2017) 46–50.

[61] M. Mishra, T.J. Callahan, T. Chelcea, G. Venkataramani, S.C. Goldstein, M. Budiu, Tartan: evaluating spatial computation for whole program execution, SIGOPS Oper. Syst. Rev. 40 (5) (2006) 163–174.

[62] H. Esmaeilzadeh, A. Sampson, L. Ceze, D. Burger, in: Neural acceleration for general-purpose approximate programs, Proceedings of International Symposium on Microarchitecture, 2012, pp. 449–460.

[63] T. Moreau, et al., in: SNNAP: approximate computing on programmable SoCs via neural acceleration, Proceedings of International Symposium on High Performance Computer Architecture (HPCA), 2015, pp. 603–614.

[64] L. McAfee, K. Olukotun, in: EMEURO: a framework for generating multi-purpose accelerators via deep learning, Proceedings of IEEE/ACM International Symposium on Code Generation and Optimization (CGO), San Francisco, CA, 2015, pp. 125–135.

[65] A.K. Singh, et al., in: Mapping on multi/many-core systems: survey of current and emerging trends, Proceedings of Design Automation Conference, 2013.

[66] Y. Hoskote, et al., A 5-GHz mesh interconnect for a Teraflops processor, IEEE Micro 27 (5) (2007) 51–61.

[67] J. Kim, J. Balfour, W. Dally, in: Flattened butterfly topology for on-chip networks, Proceedings of International Symposium on Microarchitecture, 2007, pp. 172–182.

[68] R. Kumar, A. Gordon-Ross, MACS: a highly customizable low-latency communication architecture, IEEE Trans. Parallel Distrib. Syst. 27 (1) (2016) 237–249.

[69] S. Murali, G. De Micheli, in: Bandwidth-constrained mapping of cores onto NoC architectures, Proceedings Design, Automation and Test in Europe Conference and Exhibition, 2004, pp. 896–901.

[70] M. Modarressi, H. Sarbazi-Azad, A. Tavakkol, in: An efficient dynamically reconfigurable on-chip network architecture, Proceedings of Design Automation Conference (DAC), 2010.

[71] U. Ogras, R. Marculescu, It's a small world after all: NoC performance optimization via long-range link insertion, IEEE Trans. Very Large Scale Integr. VLSI Syst. 14 (7) (2006) 639–706.

[72] H. Yang, J. Tripathi, N.E. Jerger, D. Gibson, in: Dodec: random-link, low-radix on-chip networks, Proceedings of International Symposium on Microarchitecture, Cambridge, 2014, pp. 496–508.

[73] S. Murali, et al., Synthesis of predictable networks-on-chip-based interconnect architectures for chip multiprocessors, IEEE Trans. Very Large Scale Integr. VLSI Syst. 15 (8) (2007) 869–888.

[74] J. Cong, B. Xiao, in: Optimization of interconnects between accelerators and shared memories in dark silicon, Proceedings of ICCAD, 2013, pp. 630–637.

[75] L. Yang, W. Liu, W. Jiang, M. Li, P. Chen, E. Sha, FoToNoC: a folded torus-like network-on-chip based many-core systems-on-chip in the dark silicon era, IEEE Trans. Parallel Distrib. Syst. 28 (7) (2017) 1905–1918.

[76] H. Bokhari, H. Javaid, M. Shafique, J. Henkel, S. Parameswaran, in: Malleable NoC: dark silicon inspired adaptable network-on-chip, Proceedings of Design, Automation & Test in Europe Conference & Exhibition (DATE), 2015, pp. 1245–1248.

[77] W.J. Dally, B. Towles, Principles and Practices of Interconnection Networks, Morgan Kaufmann Publishers, 2004.

[78] B. Fu, Y. Han, J. Ma, H. Li, X. Li, in: An abacus turn model for time/space-efficient reconfigurable routing, Proceedings of ISCA, 2011, pp. 259–270.

[79] P. Gratz, B. Grot, S.W. Keckler, in: Regional congestion awareness for load balance in networks-on-chip, Proceedings of International Symposium on High Performance Computer Architecture, 2008, pp. 203–214.

[80] A. Shafiee, M. Zolghadr, M. Arjomand, H. Sarbazi-Azad, in: Application-aware deadlock-free oblivious routing based on extended turn-model, Proceedings of International Conference on Computer-Aided Design (ICCAD), 2011, pp. 213–218.

[81] R. Das, O. Mutlu, T. Moscibroda, C. Das, Argia: a network-on-chip exploiting packet latency slack, IEEE Micro 31 (1) (2011) 29–41.

[82] R. Das, O. Mutlu, T. Moscibroda, C.R. Das, in: Application-aware prioritization mechanisms for on-chip networks, Proceedings of International Symposium on Microarchitecture (MICRO), 2009, pp. 280–291.

[83] J.W. Lee, M.C. Ng, K. Asanovic, in: Globally-synchronized frames for guaranteed quality-of-service in on-chip networks, Proceedings of ISCA, 2008, pp. 89–100.

[84] E. Bolotin, et al., in: The power of priority: NoC based distributed cache coherency, Proceedings of International Symposium on Networks-on-Chip, 2007, pp. 117–126.

[85] G. Michelogiannakis, et al., in: Packet chaining: efficient single-cycle allocation for on-chip networks, Proceedings of International Symposium on Microarchitecture, 2011, pp. 83–94.

[86] C.A. Nicopoulos, in: ViChaR: a dynamic virtual channel regulator for network-on-chip routers, Proceedings of International Symposium on Microarchitecture, 2006, pp. 333–346.

[87] A.K. Mishra, N. Vijaykrishnan, C.R. Das, in: A case for heterogeneous on-chip interconnects for CMPs, Proceedings of ISCA, 2011, pp. 389–400.

[88] R. Das, et al., in: Catnap: energy proportional multiple network-on-chip, Proceedings of ISCA, 2013, pp. 320–331.

[89] E. Momenzadeh, et al., in: Parallel forwarding for efficient bandwidth utilization in networks-on-chip, Proceedings of ARCS, 2017, pp. 152–163.

[90] S.H. Seyyedaghaei, A. Mazloumi, M. Modarressi, P. Lotfi-Kamran, Dynamic resource sharing for high-performance 3-D networks-on-chip, IEEE Comput. Archit. Lett. 15 (1) (2016).

[91] A. Kumar, L. Peh, N.K. Jha, in: Token flow control, Proceedings of International Symposium on Microarchitecture, 2008, pp. 342–353.

[92] M. Ahn, E.J. Kim, in: Pseudo-circuit: accelerating communication for on-chip interconnection networks, Proceedings of International Symposium on Microarchitecture, 2010, pp. 399–408.

[93] M. Modarressi, et al., VIP: virtual point-to-point connections in NoCs, IEEE Trans. Comput. Aided Des. Integr. Circuits Syst. 29 (6) (2010) 855–868.

[94] T. Krishna, et al., in: Breaking the on-chip latency barrier using smart, Proceedings International Symposium on High-Performance Computer Architecture (HPCA), 2013.

[95] P. Lotfi-Kamran, M. Modarressi, H. Sarbazi-Azad, An efficient hybrid-switched network-on-chip for chip multiprocessors, IEEE Trans. Comput. 65 (5) (2016) 1656–1662.

[96] P. Lotfi-Kamran, B. Grot, B. Falsafi, in: NOC-out: microarchitecting a scale-out processor, Proceedings of International Symposium on Microarchitecture, 2012, pp. 177–187.

[97] T. Moscibroda, O. Mutlu, in: A case for bufferless routing in on-chip networks, Proceedings of ISCA, 2009, pp. 196–207.

[98] M. Hayenga, N. Jerger, M. Lipasti, in: SCARAB: a single cycle adaptive routing and bufferless network, Proceedings of International Symposium on Microarchitecture, 2009, pp. 244–254.

[99] A. Abousamra, A.K. Jones, R. Melhem, in: Proactive circuit allocation in multiplane NoCs, Proceedings of Design Automation Conference, 2013, pp. 35:1–35:10.

[100] R.A. Stefan, A. Molnos, K. Goossens, dAElite: a TDM NoC supporting QoS, multicast, and fast connection set-up, IEEE Trans. Comput. 63 (3) (2014) 583–594.

[101] A.K. Abousamra, R.G. Melhem, A.K. Jones, in: Déjà Vu switching for multiplane NoCs, Proceedings of International Symposium on Networks-On-Chip, 2012, pp. 11–18.

[102] L. Peh, W.J. Dally, in: Flit-reservation flow control, Proceedings of International Symposium on High Performance Computer Architecture, 2000, pp. 73–84.

[103] P. Koka, et al., in: A Micro-architectural analysis of switched photonic multi-chip interconnects, Proceedings of ISCA, 2012, pp. 153–164.
[104] Y. Pan, et al., in: Firefly: illuminating future network-on-chip with nanophotonics, Proceedings of ISCA, 2009, pp. 429–440.
[105] A. Rezaei, D. Zhao, M. Daneshtalab, H. Zhou, in: Multi-objective task mapping approach for wireless NoC in dark silicon age, Proceedings of Euromicro International Conference on Parallel, Distributed and Network-based Processing, 2017, pp. 589–592.
[106] D. Park, et al., in: MIRA: a multi-layered on-chip interconnect router architecture, Proceedings ISCA, 2008, pp. 251–261.
[107] S. Jeloka, R. Das, R.G. Dreslinski, T. Mudge, D. Blaauw, in: Hi-rise: a high-radix switch for 3D integration with single-cycle arbitration, Proceedings of International Symposium on Microarchitecture, 2014, pp. 471–483
[108] BookSimNoC Simulator, http://nocs.stanford.edu/booksim.html, retrieved on Jan. 2018.
[109] C. Sun, et al., in: DSENT - a tool connecting emerging photonics with electronics for opto-electronic networks-on-chip modelling, Proceedings of International Symposium on Networks-on-Chip, 2012, pp. 201–210.
[110] SPLASH-2, http://www.flash.stanford.edu/apps/SPLASH, retrieved on Jan. 2018.
[111] K. Srinvasan, K. Chatha, in: A low complexity heuristic for design of custom network-on-chip architectures, Proceedings of Design Automation and Test in Europe (DATE), 2006, pp. 130–135.
[112] M. Schmitz, Energy Minimization Techniques for Distributed Embedded Systems, PhD Thesis, University of Southampton, 2003.
[113] D. Rahmati, et al., Computing accurate performance bounds for best effort networks-on-chip, IEEE Trans. Comput. 62 (3) (2013) 452–467.

ABOUT THE AUTHORS

Mehdi Modarressi received the B.Sc. degree in computer engineering from Amirkabir University of Technology, Tehran, Iran, in 2003, and the M.Sc. and Ph.D. degrees in computer engineering from Sharif University of Technology, Tehran, Iran, in 2005 and 2010, respectively. He is currently an assistant professor with the Department of Electrical and Computer Engineering, University of Tehran, Tehran, Iran, where he leads the Parallel and Network-based Processing Research laboratory. His research focuses on different aspects of high-performance computer architecture and parallel processing with a particular emphasis on networks-on-chip and neuromorphic architectures.

Hamid Sarbazi-Azad received his B.Sc. in electrical and computer engineering from Shahid-Beheshti University, Tehran, Iran, in 1992, his M.Sc. in computer engineering from Sharif University of Technology, Tehran, Iran, in 1994, and his Ph.D. in computing science from the University of Glasgow, Glasgow, UK, in 2002. He is currently professor of computer engineering at Sharif University of Technology and heads the School of Computer Science, Institute for Research in Fundamental Sciences (IPM), Tehran, Iran. His research interests include high-performance computer/memory architectures, NoCs and SoCs, parallel and distributed systems, performance modeling/evaluation, and storage systems, on which he has published over 300 refereed conference and journal papers. He received Khwarizmi International Award in 2006, TWAS Young Scientist Award in engineering sciences in 2007 and Sharif University Distinguished Researcher awards in years 2004, 2007, 2008, 2010, and 2013. He is now an associate editor of ACM Computing Surveys, Elsevier Computers and Electrical Engineering, and CSI Journal on Computer Science and Engineering.

CHAPTER SEVEN

Introduction to Emerging SRAM-Based FPGA Architectures in Dark Silicon Era

Zeinab Seifoori, Zahra Ebrahimi, Behnam Khaleghi, Hossein Asadi
Department of Computer Engineering, Sharif University of Technology, Tehran, Iran

Contents

Abstract

The increased leakage power of deep-nano technologies in the one hand, and exponential growth in the number of transistors in a given die particularly in *Field-Programmable Gate Arrays* (FPGAs) have resulted in an intensified rate of static power dissipation as well as power density. This ever-increasing static power consumption acts as a *power wall* to further integration of transistors and has caused the breakdown of Dennard scaling. To meet the available power budget and preclude reliability challenges associated with high power density, designers are obligated to restrict the active percentage of the chip by powering off a selective fraction of silicon die, referred to as *Dark Silicon*. Several promising architectures have been proposed to enhance the static power and energy efficiency in FPGAs. The main approach in the majority of suggested

Advances in Computers, Volume 110
ISSN 0065-2458
https://doi.org/10.1016/bs.adcom.2018.04.002

architectures includes applying power gating to unused logic and routing resources and/or designing power-efficient logic and routing elements such as *Reconfigurable Hard Logics* as an alternative for conventional *Look-up Tables*. This study represents a survey on evolution of SRAM-based FPGA architectures toward the era of dark silicon.

1. INTRODUCTION

Field-Programmable Gate Arrays (FPGAs) were first emerged four decades ago. Since their inception, these devices have experienced a rapid growth in industry and have been a viable alternative infrastructure for *Application-Specific Integrated Circuits* (ASICs). Nowadays, FPGAs have become an indispensable part of digital systems and have widespread usage in a wide domain of digital systems from embedded processors to parallel and high-performance computing to safety-critical applications. This ubiquity of FPGA usage is motivated by the intrinsic and distinctive characteristics such as flexibility to implement a variety of digital designs and adaption with workload variations, high degree of parallelism, low *Nonrecurring Engineering* cost, inexpensive design update, and short time to market.

The evolution path of FPGA had a great growth pace so that today's typical FPGA platforms, e.g., Stratix 10GX5500, comprise more than 30 billion transistors. However, such aggressive trend in transistor density necessitates the massive power and more importantly, cooling equipment so that all transistors on the chip can be employed in parallel. In recent years, however, this aggressive trend in integration of transistors in semiconductor industry has hit a *Power Wall* due to the failure of Dennard Scaling.

Dennard scaling states that transistor feature size and voltage scale commensurately by the same factor in each process generation, so the power density remains constant. This facilitates designers to increase performance by raising the clock frequency in every generation. Nonetheless, since the last decade, the reduction in supply voltage was not adequate to handle the significant increase of static power, which was the new power bottleneck. Therefore, the increasing static power and subsequently, the breakdown of Dennard scaling hinders cramming ever-more silicon into tiny areas and fully utilize the entire transistor pool on the chip within a given and safe *Thermal Design Power*. This means one may fabricate dense chips, but cannot afford the power and hence obligated to restrict the active fraction of silicon die, which is known as *dark silicon*.

When it comes to FPGAs, Dark silicon gets more pronounced, since the flexibility of FPGA devices is not only the reason of their lower performance and less area efficiency in contradiction to ASIC counterpart, but also the cause of higher static power dissipation that stems from the abundance of transistors and SRAM configuration bits aligned in logic and routing fabrics [1]. In particular, FPGA-based designs consume more static power (5–87 ×) relative to their ASIC-based counterparts [1]. This high static power dissipation obstructs their usage in low power and embedded mobile systems.

Previously, FPGA practitioners focused to enhance primarily the performance and area efficiency of FPGAs and spent quantities of transistors to achieve higher performance. However, considering importance of today's power-constrained computations and increasing rate of dark portion of silicon, they conversely spend (or waste) abundance of transistors to buy energy, managing the operating temperature, or longer battery life. With the emergence of reconfigurable devices, FPGAs have created an opportunity to be promising alternatives of processors since they tend to be more energy-efficient and outperform processors by an order-of-magnitude or even more, especially in applications that are highly parallel [2,3]. Thus, the higher power reduction of FPGAs would suggest further opportunity of their popularity.

This chapter targets the evolution of logic and routing fabrics in FPGAs toward the era of dark silicon. To this end, the rest of this chapter is organized as follows. First, an overview of SRAM-based FPGA architecture including logic and routing resources is presented in Section 2. Afterward, Section 3 provides an introduction on the concepts of power wall and dark silicon. Sections 4 and 5 summarize a few studies which offer new FPGA logic and routing architectures, respectively, targeting the mitigation of dark silicon problem. Finally, we will conclude this chapter and address important questions in the future trends.

2. ARCHITECTURE OF SRAM-BASED FPGAs

FPGAs are prefabricated reconfigurable devices that can be programed to implement various digital systems. The flexibility of implementing versatile applications is facilitated by simply loading the target bitstream and reconfiguring the FPGA SRAM configuration bits, divided into logic and routing fabrics. The logic blocks are responsible to implement the application functionality, while routing resources provide the communication among logic blocks. The arrangement of logic and routing network determines the FPGA topology and therefore the overall efficiency of the FPGA.

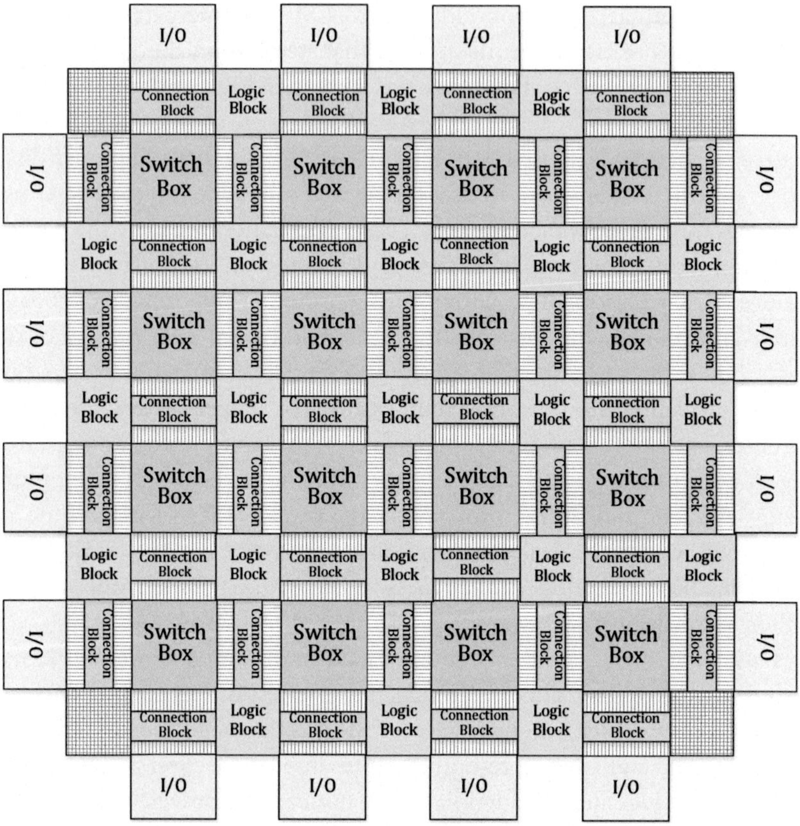

Fig. 1 The overall architecture of island-style FPGAs.

A typical architecture of FPGA that is also preferred by most FPGA commercial vendors (such as Altera and Xilinx) is island-style which is illustrated in Fig. 1. In the island-style architecture, logic blocks are arranged as a two-dimensional array within a pool of routing resources.

2.1 Configurable Logic Block

Logic resources are the main computational components of FPGAs that implement and store the functionality of the target circuit. Logic resources are composed of a two-dimensional matrix of *Configurable Logic Blocks* (CLB) that are also simply referred as clusters. As shown in Fig. 2, each CLB is a cluster of *Basic Logic Element* (BLE)—typically between 4 and 10—and intracluster multiplexers that conduct the routing channel signals toward the BLEs. Logic architecture can be either *homogeneous* or *heterogeneous*.

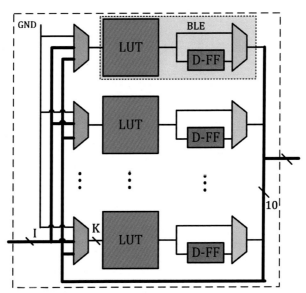

Fig. 2 The structure of a Configurable Logic Block (CLB).

In the former, all CLBs consist of a similar BLE structure, while in the latter case, heterogeneity can be intracluster or intercluster. Intracluster heterogeneity is identified by a mixture of dissimilar BLEs inside a cluster, while intercluster heterogeneousness means each column of CLBs is dedicated to a certain type of BLEs.

Each BLE is mainly composed of three elements. First, a basic functional block which has been provided with a range of choices such as K-input *Look-up Table* (LUT), *Reconfigurable Hard Logic* (RHL), or *Programmable Array Logic* (PAL) style logic gates. Second, a storage element such as flip-flop which is crucial for sequential designs. Third, a 2-MUX that selects either the output LUT or flip-flop on the structure of the target circuit.

Throughout the history of FPGAs, LUTs have been the mainstream logic block adopted by commercial FPGA vendors. Each K-input LUT can implement any function with up to K variables and is comprised of (a) 2^K SRAM configuration bits that store the truth table of target Boolean function and (b) 2^K multiplexer that selects the appropriate SRAM content based on the given inputs of the LUT. The structure of 4-LUT is demonstrated by Fig. 3 [4].

Various structures (i.e., number of inputs) of LUT enable a spectrum of area, power, and performance trade-off in FPGA designs. Previous studies

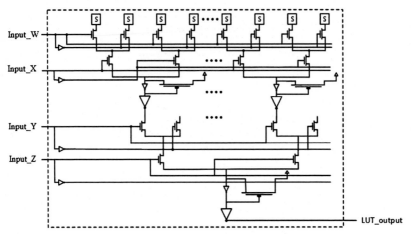

Fig. 3 The typical structure of 4-LUT [4].

have consistently shown that 4–LUT-based FPGAs are well-suited for area efficiency, while 6–LUT-based FPGAs provide the highest performance among different LUT structures but sacrifices the area, as a consequent [5]. Although large-input LUTs such as 6-LUT absorb more logic and subsequently reduce the critical path length, the number of SRAM cells and LUT transistors increases exponentially as the number of LUT inputs increases.

2.2 Hard-Core Blocks

Many new FPGA devices have been equipped with various primitives such as memory blocks (single/dual port RAMs), fracturable multipliers, and *Digital Signal Processors* (DSPs). These DSP and other hard-core blocks are optimally designed and embedded in the devices to facilitate implementing specific functions that otherwise require much larger number of LUTs to be implemented and provide an opportunity to implement applications with memory demand. Nonetheless, architectural parameters, e.g., repetition frequency of the column blocks, as shown in Fig. 4 [6], are a key design parameter and influence the performance as well as energy efficiency of FPGAs. For instance, previous studies have shown frequency of 5 and 12—putting a column of hard-core blocks in every 5 and 12 columns—provides the optimum area and energy efficiency when such molecules are prevalent and rarely used in the design, respectively [7,8]. However, tuning the size of these blocks is still under development.

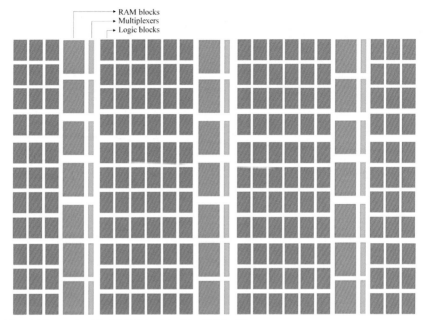

Fig. 4 The embedded DSP block inside FPGAs [6].

2.3 Routing Resources

The routing fabric provides the connectivity among a pool of programmable logic blocks. The interconnection network contains configurable switch matrices and connections blocks that can be programmed to form the demanded connection. In the island-style architecture, routing resources are divided into *Connection Blocks* (CBs) and a matrix of *Switch Boxes* (SBs), namely, *Switch Matrix* (SM). CBs are responsible to provide connection between logic cluster input/output pins to the adjacent routing channels. The portion of routing channels that is connected to the logic cluster input is denoted by F_{cin}. Analogously, the portion that determines the connectivity of logic cluster output pins to the adjacent routing channels is termed as F_{cout}. The structure of a CB with $F_{cin} = 0.5$ is illustrated in Fig. 5.

SBs are placed at the intersection points of vertical and horizontal routing channels. Routing a net from a CLB source to the target CLB sink necessitates passing through multiple tracks and SBs, in which an entering signal from a certain side can connect to any of the other three directions based on the SM topology. The popular SB topologies in commercial FPGA architectures are *Wilton*, *Disjoint*, and *Universal* which are shown in Fig. 6. The parameter F_s is the flexibility of the SB which indicates the number of output

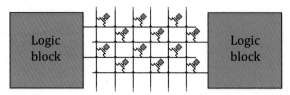

Fig. 5 The structure of connection block with Fc = 0.5.

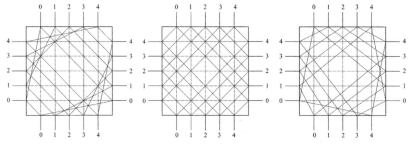

Fig. 6 Three prevalent SM topology—Disjoint (*left*), Universal (*middle*), and Wilton (*right*).

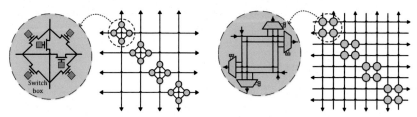

Fig. 7 The structure of uni (*left*) and bi (*right*) Universal SM.

branches of a track entering an SB. F_s is usually set to 3. Interconnections channel can be *unidirectional* and/or *bidirectional* as illustrated in Fig. 7. Modern FPGAs, however, are mainly consisted of unidirectional tracks.

The routing tracks can have single- or multisegment length, e.g., a wire segment with length of two spans two CLBs. Longer wires encounter fewer SB (multiplexer) delays but have longer length and larger delay and are optimal for global routings, while tracks with shorter lengths have smaller wire delay and are proper for nearby routings.

3. POWER WALL AND DARK SILICON

Although transistor density has intended to grow on an exponential trajectory, challenges associated with further shrinking the physical dimensions and powering on the total resources in the chip has acted as an obstacle particularly because air-cooling of the chips has remained almost unchanged or experienced little improvements over technologies. Today's cooling technology stands at the point that it can cool down the heat of a chip with the power of 175 W to keep it in the safe temperature at 85°C [9]. This is while developing an exascale computer system in today's technology requires about 3% of a typical nuclear plant's generating power that makes it somehow impractical due to the excessive power requirement [10]. This problem can threaten the future perspective of hardware, especially FPGAs.

When it comes to FPGAs, the role of power dissipation, especially static power, has become more challenging and as a major design constraint. The static portion of power contributes only 22% of total power in 90 nm FPGAs, while in recent 28 nm FPGAs this is passed over 50% [11,12]. Eq. 1 demonstrates the relation between static and dynamic portions in total power [13]. In this equation, α is the switched off (dark silicon) portion, N is the number of transistors, I_{leak} is the leakage current, K is an architectural constant parameter, C is the transistor capacity, and f is the operation frequency of the chip. To reduce operation voltage V_{dd}, V_{th} should be reduced accordingly, while reducing V_{th} leads to exponential increase of I_{leak} and total power consumption. The tight dependency between threshold voltage and leakage power has become more pronounced as magnitude of V_{th} has reduced. Therefore, transistor count and operational voltage have not stuck together in line with a same upward slope, causing the upward performance trend to halt and a sharp increase in power consumption, in particular static power.

$$
\begin{aligned}
P &= P_{static} + P_{dynamic} \\
&= \alpha \times N \times I_{leak} \times V_{dd} \times K + (1 - \alpha) \times N \times C \times f \times V_{dd}^2
\end{aligned} \tag{1}
$$

Due to the downscaling of threshold voltage and transistor feature size in the last decades, designers are faced with a sharp increase in power density with each technology scaling. Fig. 8 [14] illustrates the increase of power density, and subsequently, the end of Dennard scaling according to 2013 ITRS projections [15] and Bokar conservative projections [16]. The main source of static power consumption in 90 nm technology and beyond originates from leaking the current from gate into substrate, especially in idle periods of the application status [17]. This can be translated literally as consuming power for doing nothing.

Because of the ever-increasing static power across technology scaling, several issues have been emerged such as degradation of the overall

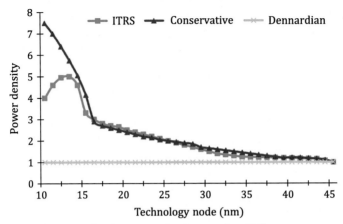

Fig. 8 Increase in power density over technology scaling [14].

Power per Operation. In addition, full utilization of resources consumes more power and raises the chip temperature, which consequently consumes higher leakage. This phenomenon is known as *Positive-Temperature-Leakage-Feedback* which can result in reliability degradations such as accelerated aging process and thermal runaway, and hence, circuit malfunction.

In order to resolve the aforementioned problems in the dark silicon era, while chips will be equipped with abundance of transistor, only a part of their silicon die can be simultaneously *powered on* and operate at high-performance condition with nominal voltage. Other portion, which may contribute a considerable amount of silicon die, is switched off, i.e., power gated to avoid consuming leakage power and making thermal hotspots in order to keep the chip temperature below the safe thermal conditions. Based on the analytical estimation in Ref. [18], it has been predicted that in 22 nm technology, the dark percentage of total chip area will exceed 50%. However, by exploiting several circuit- and system-level techniques such as exploiting *high-k* dielectric materials to reduce the leakage, and *Dynamic Voltage and Frequency Scaling* (DVFS) to reduce the dynamic power, predicted trend and estimations in dark fractions of silicon has slightly mitigated in semiconductor integrated circuits. Nevertheless, by analyzing data from Intel and ITRS projections, it is anticipated that 50%–80% of total chip area in both GPU- and CPU-based systems using 8 nm technology will be either dark or under-clocked. Considering the falling trend of utilization by a factor of $2\times$ per process generation, the dark portion is predicted to be up to 90% in 2020. The trend of dark silicon is illustrated in Fig. 9 [8,15].

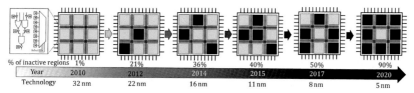

% of inactive regions	1%	21%	36%	40%	50%	90%
Year	2010	2012	2014	2015	2017	2020
Technology	32 nm	22 nm	16 nm	11 nm	8 nm	5 nm

Fig. 9 Growth of dark silicon percentage with technology scaling due to increasing static power.

While approaching the power wall, the pace of performance improvements (i.e., increasing the clock rate) with technology scaling has logged. In particular, the transistors scale by the factor of X^2, while frequency scales by X. Consequently, processor manufacturers turned to multicore approach to satisfy the high performance demands. However, this is not an ultimate solution for dark silicon since duplicating the number of cores also scales with the same rate as either dark silicon percentage or transistor shrinking factor [18]. In Ref. [19], four promising directions—four horsemen—have been suggested to cope with the power wall and to address the problem of dark silicon as follows.

1. *The shrinking horseman*: Shrink down chips in order to reduce total static power and maintain performance efficiency.
2. *The dim horseman*: Use dim and sprinting operational modes such as temporal and spatial (but brief) boost of chip resources, *Coarse-Grained Reconfigurable* Arrays (CGRAs) to reduce the multiplexing of processor datapaths, near-threshold voltage processing to diminish dynamic power, and exploit dark fractions of larger caches [20–22].
3. *The specialized horseman*: Exploit architectural heterogeneity such as application-specific accelerators and specialized coprocessors with different power and performance characteristics [18,23–26].
4. *The deus ex machina horseman*: Create novel physical materials such as low leakage post-MOSFET-based transistors, e.g., *Tunnel Field Effect Transistor* (TFET) which is based on tunneling effects and *Nano-Electro-Mechanical* (NEM) switches, high-k metal gate, nonvolatile memory cells, and strained silicon. Unfortunately, they fail in inherent privilege of easy scaling of MOSFET [27–29].

The main question posed for architectural designer and hardware/software codesign community is that: "Can an abundance of transistors in the chip be utilized to improve design metrics such as performance and power efficiency in the era of dark silicon?" If the answer is yes, then how? During the past decade, multitude of studies has been explored the phenomenon, challenges, and implication of dark silicon in many-core *System on-Chips* (SOCs) and FPGAs. In these studies, various approaches or architectures have been proposed to throttle the static power and/or enhance the thermal management and reliability. Moreover, various architectures have been proposed to harness the dark fractions of the chip to improve the desired objective within the given power envelope such as mitigating aging and manifesting soft-errors, process variations, and reliability concerns [30–34]. In the following, we focus on few studies that target enhancement of power efficiency in FPGA logic and routing fabrics.

4. LOGIC BLOCK ARCHITECTURES IN DARK SILICON ERA

The logic resources consume significant portion (about 50%) of the static power in SRAM-based FPGAs fabricated in 28 nm and beyond [11,12]. In most recent studies, various logic architectures have been proposed as a promising solution to effectively accommodate with the dark silicon trend. These studies leverage abundance of transistors in such a way that enhances the power efficiency of the FPGAs as well as preserving the performance [8,11,12,35–37]. Inspired by coarse-grained heterogeneity solution for dark silicon problem in many-core systems, fine-grained heterogeneity has also recently started to find a foothold in FPGA logic resources.

In the logic, the heterogeneity of hard logics along with conventional soft logics has been explored to improve power efficiency as well as the performance of the FPGAs. The main incentive behind the presented studies is laid down under the exploitation of RHLs in fine-grained heterogeneous architecture and/or power gating of unused logic resources. RHLs are a set of logic gates with fewer configuration cells as compared to their LUT counterparts and can implement a significant portion of Boolean functions with considerably smaller area, power, and delay as well as less susceptibility to energetic particle strikes because of fewer number of SRAMs.

Power gating can be applied static or dynamic; the former is applied offline at the configuration time, while the latter is applied online during the application runtime. The power efficiency of static technique is limited to the permanently off regions. In contrary, dynamic power gating may have higher opportunity for power saving since each component can be temporarily switched off in its idle periods. In practice, however, dynamic power gating encounters several difficulties. First, the most challenging issue with dynamic power gating is inrush current, i.e., a substantial wake-up current drawn from power supply whenever a power gated component needs to be powered on. This phenomenon can cause wake-up time penalty (and timing violations if did not properly predicted) and data instability in flip-flops that eventuates in functional errors [12]. Second, dynamic power gating requires complicated power controller and techniques to (a) predict the idle periods and estimate whether the power gating is advantageous despite the related overheads, i.e., wake-up transitions energy, power controller energy, and overhead of routing the controlling signal; (b) anticipate the wake-up time in order to initiate the powering on phase to prevent timing and functional errors which needs complicated techniques in environment interactive and time-dependent applications while assuring that only a small portion of resources are powered on each time to offset inrush current; and (c) route the power controlling signals to each unit which may complicate the overall routing. In the following, we elaborate several logic block architecture with the aim of reducing the power consumption. We categorize these studies as architectures with logic building blocks similar to conventional FPGAs, and architectures that propose novel logic elements. Notice that there are a set of studies that target reducing the power of both logic and routing resources. We will elaborate these joint studies while discussing the interconnection network.

4.1 Low-Power FPGA With Conventional Logic Architecture

Ahmed et al. [37] have proposed an approach for detecting dynamic power gating opportunities in a reconfigurable SoC framework comprising a MIPS processor connected to an accelerator which are generated by *High-Level Synthesis* (HLS) methodology. The HLS scheduling information automatically identifies the idle intervals of accelerator by application profiling and extracts the power-state controller, as well. The suggested approach focuses on accelerator-level opportunities which means when the entire accelerator will be switched off, power gating is promising. This approach is profitable when the accelerator is exploited for a specific part of the application such as loop unrolling in which operations can be executed in parallel. The experimental results over a set of CHStone benchmarks that are exploited for C-based HLS has shown the suggested approach achieves considerable leakage energy savings (up to 96%) in five (out of 12) applications. However, details such as power on energy and time overhead and inrush current issue are neglected in this work.

In Refs. [11,12], Bsoul et al. have proposed an architecture that supports static and dynamic power gating. In the suggested architecture, an FPGA is divided into several square-shaped regions wherein each one is composed of customizable, but predetermined number of CLBs. In each region, all CLBs and adjacent CBs are configured as either three modes: regions that are frequently used are configured to always-on, while unused ones are in always-off state. Conversely, regions that are idle for a long interval are dynamically power gated during run-time by using signals that are generated and routed from a power controller. Thus, all SBs throughout the FPGA need to be always-on state to route the controlling signals. The overall structure of the suggested architecture is shown in Fig. 10. The experimental results over a set of MCNC benchmarks and a robot control system which is used in medical applications has demonstrated considerable leakage energy saving (up to 83%) with affordable overhead (about 10%) in area and performance.

In Ref. [36], Ishihara et al. have proposed an asynchronous FPGA with autonomous fine-grained power gating. The structure of the proposed architecture is a mesh array of logic blocks (LBs) consisting of 2-input LUTs, a power controller, and an SB which operates in three modes, i.e., *sleep*, *standby*, and *active*. The suggested architecture requires four additional wires: two for data, one for acknowledge, and another one for wake-up that is used to wake-up the next LB before being accessed. In the asynchronous architecture, the activity of a power-gated LB is detected by comparing the phases of the input data with that of the output data. If the phases of the input and output are the same, then the LB is idle and can be power gated if it will not be requested for at least 10 ns. To alleviate the latency of dynamic power gating, not only the demanded LB will be switched on, but also the phase of next LB will be changed to standby. Experimental results over four tunable benchmarks have shown 15% static power reduction in the suggested architecture. However, being limited to 2-LUTs imposes significant area and dynamic power overhead to the suggested architecture. Figs. 11 and 12 show the architecture and structure of the LB proposed in Ref. [36].

4.2 Low-Power FPGAs With Novel Logic Architectures

In Ref. [35], Ahari et al. have proposed a reconfigurable architecture that aims to enhance the power efficiency of FPGAs by exploiting power-efficient *Generic Reconfigurable Hard Logics* (GRHLs) as a substitute for conventional LUTs and enabling power gating of unused logic blocks (LBs). In their suggested architecture, each LB is replaced with a *Mega Cell* (MC) which consists of two GRHLs and a 4-LUT. All GRHLs and the LUT are controlled by the means of a *Power-Controlling Configuration Bits*

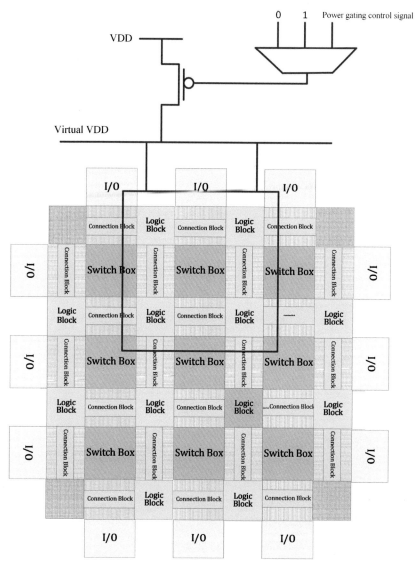

Fig. 10 The overall structure of suggested architecture in Refs. [11,12].

(PCCBs) in such a way that at most one of GRHLs/LUT that implements the function is activated at a time, and the unused ones, or the entire MC (in the case of not being used) will be power gated.

GRHLs are small generic cells that benefit from fewer SRAMs, smaller area and delay as compared to 4-LUTs. They are designed in such a way that they can cover more than 80% of the 4-input functions. The motivation

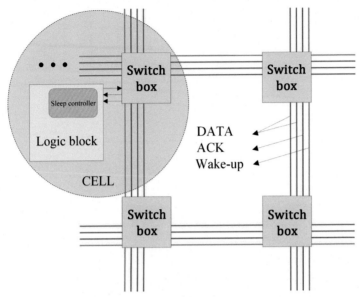

Fig. 11 The overall structure of suggested architecture [36].

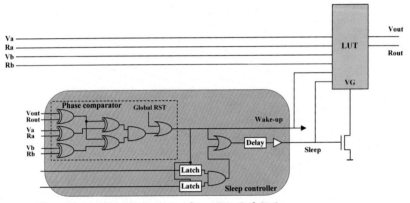

Fig. 12 The suggested block diagram of an LB in Ref. [36].

behind proposing the GRHLs are the nonuniform distribution of Boolean functions in applications. The similarity of Boolean functions is determined based on *Negation-Permutation-Negation* (NPN)-Class representation. For instance, two functions $F = A\overline{B} + CD$ and $G = DB + AC$ are NPN–equivalent since each function can be derived from the other by negating B and permuting A and D. Fig. 13 illustrates the two suggested GRHLs and Table 1 summarizes

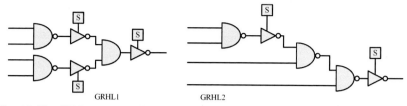

GRHL1   GRHL2

Fig. 13 The GRHLs suggested in Ref. [35].

Table 1 The Coverage Ratio of NPN-Class by Suggested GRHLs in [35]

NPN Classes	ABCD	A(B+CD)	AB(C+D)	AB+CD	A(B+C+D)	Other	Total
GRHL1	✓		✓	✓			
GRHL2		✓			✓		
Ratio	37.1	8.7	12.1	16.1	9.6	16.4	100

their coverage ratio of 4-input NPN classes in 20 MCNC benchmarks. As summarized in this table, GRHLs can cover 83.6% of 4-input functions. Hence, 4-LUTs are needed to implement about 16% of functions.

The overall structure of suggested MC is demonstrated in Fig. 14. As shown in this figure, the proposed PCCB is composed of three exclusive controlling SRAMs, each of which controls the power gate state of GRHLs and LUT. Experimental results have shown their suggested architecture is able to reduce the static power, critical path delay, and *Power Delay Product* (PDP) of logic resources by 30.4%, 5.3%, and 28.8%, respectively. However, this architecture incurs an area overhead of 17%.

In Ref. [8], Ebrahimi et al. have proposed a fine-grained heterogeneous architecture, namely, PEAF that aims to improve both power and performance efficiency of SRAM-based FPGAs by employing static power gating. In PEAF, a *Reconfigurable Logic Unit* (RLU) is proposed to replace the conventional LUT-based BLE. Each suggested BLE includes three RHLs and a 3-LUT. The applicability of the proposed RHL is motivated by comprehensive function characterization in a wide range of academic and industrial benchmarks, i.e., MCNC, IWLS'05, and VTR. The investigation has revealed that up to 97% of Boolean functions have the same homomorphic structure and can be implemented with efficient RHLs and 3-LUTs. The incentive behind exploiting 3-LUT instead of 4-LUTs is the relatively high appearance of 3-input functions in the remainder portion of functions which

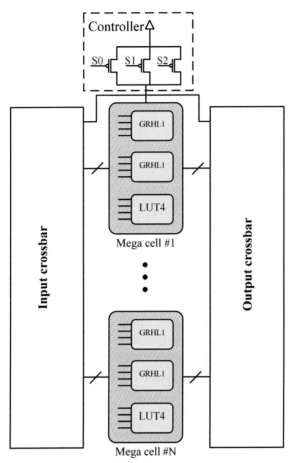

Fig. 14 The overall structure of suggested MC in Ref. [35].

cannot be implemented with suggested RHLs. The proposed RHLs and their coverage ratio are presented in Fig. 15 and Table 2, respectively.

The overall structure of the suggested RLU is illustrated in Fig. 16. Each RLU also includes a *Reconfigurable Power-Controller* (RPC) to power gate unused RHLs/LUT or the whole RLU if none of its building elements was used. A novel *Logic SRAM Sharing* (LSS) scheme has been also proposed to alleviate the area overhead in the suggested architecture. In this technique, eight SRAMs in the proposed RLU can be shared among all RHLs and 3-LUT, since at most one of them is activated at each time. Other advantages of LSS scheme are static power saving and reliability improvement of logic

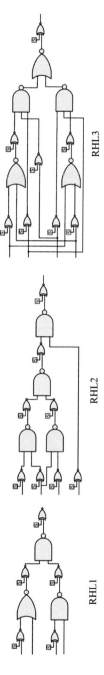

Fig. 15 The RHLs suggested in Ref. [8].

Table 2 The Coverage Ratio of NPN-Class by Suggested GRHLs in [8]

NPN Classes	Average Ratio	RHL1	RHL2	RHL3	3-LUT
ABCD	34.0	✖			
AB(C+D)	14.4	✖			
AB+CD	13.5	✖			
A(!BC+BD)	8.6		✖		
A(B+CD)	7.0		✖		
A(B+C+D)	5.1		✖		
ABCD+!(AB)!CD	4.4			✖	
AB+AC+BC	1.6				✖
A(B⊕C)	1.2		✖		
AB (C⊕**D**)	1.0			✖	
9					
A⊕**B**⊕**C**	0.9				✖
A(B⊕**C**+D)	0.9			✖	
A(BC+BD+CD)	0.6			✖	
!A!B!C+ABC	0.5			✖	✖
AB(C+D)+!A!B!C!D	1.6 (shared)			✖	
ABC+!B!C				✖	✖
(A⊕**B**)+**CD**				✖	
A(B!CD+!BC!D)				✖	
ABCD+!A!B!C!D				✖	
!A!BCD+(A⊕**B**)!**C**!**D**				✖	
ABC+!A!B!C+BC!D+!B!CD				✖	
Other supported NPNs				✖	
Unsupported NPNs	4.5				

resources which are proportional to number of SRAMs. Experimental results reveal that total static power—considering logic and routing—are improved by 24.5%. In addition, dynamic power, critical path delay, and overall PDP are also enhanced by 9.8%, 1.8%, and 21.7%, respectively.

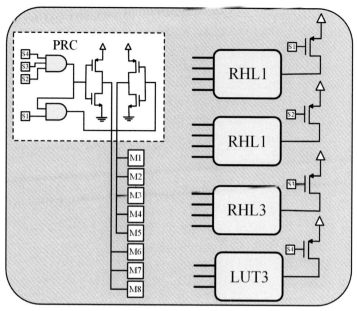

Fig. 16 The overall structure of suggested RLU in Ref. [8].

5. ROUTING BLOCK ARCHITECTURES IN DARK SILICON ERA

Previous studies to mitigate the static power consumption of routing network in FPGAs can be classified into two categories. The first category employs power gating techniques to reduce static power and the other one belongs to nonpower gating techniques. In the following, we elaborate the aforementioned studies to cope with high static power consumption of FPGAs.

5.1 Nonpower Gating Routing Architectures

Here, we investigate the nonpower gating techniques used to reduce the static power consumption of interconnection network. In this regard, Anderson et al. propose a novel architecture of FPGA routing switches that aim to overcome the high power consumption of routing fabrics employing *dual-vdd* technique [38]. The proposed structure of routing switch requires minor changes to a conventional switch and can operate in three modes of *sleep*, *low-*, and *high-speed*, each of which has different power and performance tradeoff. In low-power modes, performance is sacrificed for reducing the power consumption. This architecture employs a unique supply voltage rail and the other (than nominal) voltage level is generated internally by adding configuration SRAM cells and extra transistors to facilitate dual-Vdd technique. Indeed, this architecture takes advantage from the available timing slack of noncritical paths in designs to slow down the corresponding routing switches and reduce their power. The authors also propose a novel buffer design that can operate in different power and performance tradeoffs. As Fig. 17 illustrates, in the suggested buffer design, the body of PMOS transistors are connected to lower voltage, which leads to lower power consumption due to the body effect. Dynamic power consumption of this architecture is reduced by 28%–32% in low-speed mode compared to the high-speed one, and static power consumption is diminished by up to 52% and 79% in low-speed and sleep modes, respectively.

Krishnan et al. exploit dual-threshold technique to propose new switch block architecture to optimize the power consumption [39]. The authors replace the conventional NMOS-based pass transistor switches with variations of *Dual Threshold MOS* (DTMOS)-based switches. When the source-bulk voltage of NMOS transistor is positive, the threshold voltage decreases, and conversely, when the source-bulk voltage is negative, the threshold voltage increases. The proposed architecture takes advantage of this principle to design novel low power switch block architecture. The DTMOS transistor can be used in three forms of basic, with limiting transistor, and with augmenting transistor (see Fig. 18). In the limiting form of DTMOS, low swing voltage problem of the basic form is resolved. The authors use different variations of the basic DTMOS switch to find the best architecture in term of power consumption. The main limiting and augmenting transistors of DTMOS transistors in the proposed switches can be high or low

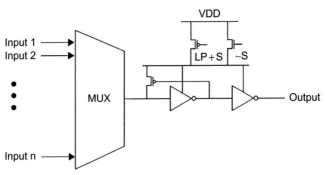

Fig. 17 Programmable low-power routing switch in Ref. [38].

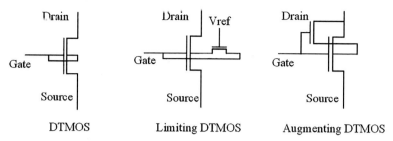

DTMOS   Limiting DTMOS   Augmenting DTMOS

Fig. 18 DTMOS with limiting and augmenting switch in Ref. [39].

threshold, while a sleep transistor in the buffer of switch box can be used, as well. Krishnan et al. found that the power consumption of *limiting switch* with high threshold main and sleep transistors is the least and *limiting switch* with high threshold main transistor and *limiting switch* are the fastest switches. Furthermore, the *limiting switch* with high threshold main transistor has the minimum power-delay product, while augmenting switch with high threshold main and sleep transistor and complementary pass transistor switch has the highest power-delay product. The proposed different variations of NMOS switches reduce the power-delay product up to 16%. The suitable architecture can be opted based on different applications requirements.

Xilinx uses triple–oxide transistors (i.e., transistors with thin, medium, or thick oxide) to reduce the power of resources that do no reside in critical path of the design [40]. The transistors have a combination of different channel lengths and threshold voltages to achieve desired performance and power consumption. Thick oxide transistors have lowest static power consumption and lower performance, as well. The reduction of static power consumption obtained by using medium oxide transistors, compared to using thin–oxide transistors, is significant while the performance degradation is not considerable. Transistors with various oxide thicknesses are employed in different FPGA resources based on their performance requirements. Xilinx also adjusts the threshold voltage of transistors through applying the back-bias voltage during the runtime to reduce the static power consumption at the cost of performance degradation. Therefore, transistors with higher performance (and higher power consumption) are used in time-critical blocks and transistors with lower performance and lower power consumption are used for unused blocks and regions that are not in the critical path.

A few studies have dealt with static power reduction by adjusting multiplexer inputs. The main incentive behind these studies is based on the dependency of static power consumption of routing multiplexers to logical value (i.e., zero or one) of the inputs. In Ref. [41], Anderson et al. have suggested two approaches to combat static power consumption. In the first approach, they change the logic signals polarities to put FPGAs resources into a low-leakage state without affecting the area and delay. In the second approach, they propose a leakage aware routing whose aim is to maximize the use of resources with less static power consumption. Using SPICE simulations reveals that the static power consumption of multiplexers depends on their input signals states. Experimental results show that when all input signals are logical "1," the static power consumption of multiplexers is minimum. In addition, when the multiplexer output is logical "1," the static power consumption is less as compared to logic "0." The static power consumption of buffers for both input states (logic "0" or logic "1") is almost the same but the static power consumption of multiplexers in logic state "0" is larger by 20%, as compared to logic state "1." As shown in Fig. 19, to enable signals to spend most of their time in a low-leakage state (i.e., "1"), the "0" signals should be inverted through inverting the configuration bits of LUTs which necessitates permuting the configurations of fanout LUT(s), accordingly. Due to the fact that various FPGA routing switch types (i.e., different multiplexer sizes) have different static power consumption, in the second approach, the authors modify the routing algorithm in a manner that exploits paths with less leaky resources between two nodes. It adapts routing by attributing a power-dependent cost to routing resources. Experimental results show that the first approach reduces

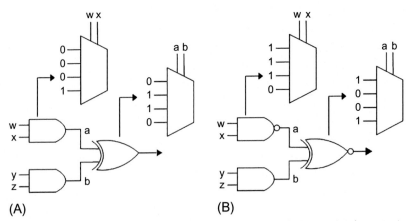

Fig. 19 LUT circuit implementation; signal inversion. (A) Circuit implementation. (B) Signal inversion in Ref. [8].

the static power by 25% on average, and by using both approaches jointly, the static power consumption decreases up to 44%.

Srinvasan et al. have focused on static power consumption of unused multiplexers by manipulating their input signals [42]. They found that the static power consumption of multiplexers depends to the logic state of multiplexers input signals, and consequently, they attempt to place the multiplexers in their low-leakage states by changing the value of unused inputs. Particularly, as the number of "0" inputs of multiplexers increases, it consumes lower leakage power. To control the logical value of multiplexer inputs, this study augments the multiplexers output by adding a configurable circuitry to assign the desired (i.e., power-friendly) value to the output of unused multiplexers of the previous stage (i.e., the driving multiplexers). Due to the high percentage of unused multiplexers in routing networks (up to 9%), this approach can reduce the static power of routing multiplexers and the overall static power by 70% and 38%, respectively. This study, however, does not consider the large output buffers of multiplexers which may consume higher power by passing "0" and conserve the power reduction achieved from the multiplexers.

In a similar approach, Hasan et al. [43] consider both multiplexer as well as its associated output buffer for achieving the minimum leakage power consumption by controlling the multiplexers input signals. They conversely find out that setting all inputs of multiplexers to logical "1" consumes less power than setting them to "0." However, no scheme has been proposed in this regard. It should be noted that both [42,43] assume simple multiplexer and buffer structure without respecting the actual transistor sizing of commercial devices which may adverse the obtained results.

5.2 Power Gating Routing Architectures

There are several prior studies that aim to reduce static power through power gating with different levels of fine- or coarse-grained each of which can be implemented statically and dynamically. Due to the undesirable area overhead of fine-grained power gating, for applications with great amount of idle resources, coarse-grained power gating is more effective. In statically controlled power gating, the on/off state of each power gating region of FPGA is determined during the configuration time of the design and this state does not change in the runtime since the resources are permanently unused. Hence, static power gating is more effective for FPGA designs with a large amount of unused resources. Dynamically controlled power gating provides the possibility of turning off the temporarily inactive resources during the runtime. Therefore, the power reduction opportunity obtained by dynamically controlled power gating is more than statically controlled power gating. Nevertheless, implementing dynamically controlled power gating, especially in routing resources, is far complicated than the statically controlled power gating. Related studies on power gating for FPGA routing resources are detailed in the following.

Bsuol et al. employ dynamically controlled power gating for reducing the static power consumption of logic blocks [44]. Despite the significant contribution of SB multiplexers in the static power consumption of FPGAs, the authors assume that SBs are always on to be able to route the power gating control signals.

Bharadwaj et al. reduce the static power through grouping the FPGAs resources and switching them off in the idle periods by a *Power State Controller* (PSC) [45]. This study extracts the idleness periods of applications from their *Data Flow Graph* (DFG) and clusters the components with similar idleness periods in the same group. The authors have conducted the experiments with a different number of clusters and investigated the effect of the number of clusters on energy, area, and power consumption. Their results indicate that the minimum power consumption and energy can be achieved at the expense of higher area overhead.

Gayasen et al. cope with static power consumption using a dynamic fine-grained power gating architecture [46]. The authors also propose a placement algorithm, namely, *Region Constrained Placement* (RCP), to prevent scattered placement of logic blocks and to increase the number of unused regions that can be power gated. Notice that in CLB-level power gating wherein each CLB is controlled by a single power switch, the notion of RCP does not apply. However, CLB-level power gating incurs considerable area overhead, and hence, [46] targets regions with large number of CLBs. RCP algorithm is implemented in continuous horizontal and vertical placement forms. In horizontal RCP, CLBs are utilized (filled) row by row while the vertical RCP fills the columns consecutively. Despite the same number of assigned regions in both cases, due to the asymmetry of FPGA array, these approaches represent different performance. For the sake of routability, the signals of used regions can also utilize routing resources outside the region. The proposed time-based control scheme also dynamically turns off the regions whose resources are completely idle. The authors combine the time-based control with different RCP variations, namely, module-level RCP that places each module of the design with a unique idleness behavior, and design-level RCP that places the whole design with placement constraints. Experiment results show that module-level RCP is the most efficient placement algorithm while design-level RCP also outperforms the typical placement algorithm most of the time.

Li et al. combat static power consumption through dynamic coarse-grained power gating and propose a placement algorithm based on sleep regions [47]. The power gating regions of the proposed architecture comprise CLB and CB. The backbone of their proposed algorithm is a power gating circuit, namely, *Power Controller Hard Macro* (PCHM) which controls the state of power gating regions and its clock. They also adapt the placement algorithm by varying the cost function in such a way that the number of idle regions increases, thus, lower power consumption is incurred. In spite of a significant fraction of static power of SBs, it is neglected in this study.

Bsoul et al. propose new architecture to dynamically power gate SBs and LBs in idle periods [48]. The authors also modify the CAD flow to increase the opportunities of power gating. The granularity of the suggested power gating architecture is at whole SM level and an SM can be turned off only if all SBs belonging to a particular SM are idle. However, several SBs should remain on for routing the control signals even if the adjacent logic blocks are powered off. FPGA resources in the proposed architecture can be in one of three states of *always-on*, *always-off*, and *power-controlled*. This study controls the SBs power gating statically, that is, SBs can only be either in *always-on* or *always-off* state. Those SBs that are responsible for the route control signals and nets of the logic blocks are always-on and the remaining unused SBs are powered off. Placing the logic blocks in a bounded area increases the number of the inactive SBs, and hence, increases the power gating opportunity. In the suggested placement algorithm, the design is confined to be placed in a specific rectangular area. The authors also optimize the routing by modifying the congestion term of nodes cost to avoid using SBs outside the bounded area of the module. Investigating power gating regions with different sizes concludes that larger region sizes lead to more power saving. This is due to higher number of control signals and always-on SBs to route controlling signals in the case smaller power gating regions are used. Despite the large number of unused multiplexers in always-on SBs, the proposed architecture does not examine the opportunity of partially power gating such SBs.

In Ref. [49], Yazdanshenas et al. have proposed a fine-grained static power gating approach which aims to reduce the power of both logic and routing networks. For the logic resources, the proposed architecture profits from the fact that the majority of LUTs in the FPGA-mapped designs are either unused or partially used. That is, in a 6-LUT architecture, a considerable portion of used LUTs use five or fewer number of inputs. Accordingly, considering the multiplexer-based structure of LUT in which a 6-input LUT is composed of two 5-input LUTs or four 4-input LUTs (and so on), the authors subdivide each 6-LUT to smaller ones and cut-off the unused sub-LUTs using an associated power gating transistor and controlling SRAM. There is also a trade-off between the number of sub-LUTs and power gating area; higher number of sub-LUTs (i.e., using smaller sub-LUTs such as 3-LUT) increases the opportunity of power gating at the cost of increased area overhead. The authors assume a fully pass transistor-based CB structure (which is inefficient and is not employed in commercial FPGAs) wherein an N-input CB multiplexer is employed as N pass transistors each of which connected to a routing track by a configuration cell while at most one of them is active in each design (to route the appropriate track to LB input). Therefore, the authors classify these N-inputs into several subgroups and turn off the unused subgroups, i.e., all of the subgroups except the one in which the selected (used) input resides in. For SBs, the authors suppose a simple bidirectional pass transistor based one which comprises six configuration SRAM cells to enable the appropriate path(s). By exploring the frequency of used patterns, the authors realize that choosing the granularity of one cut-off transistor for a whole SB achieves the best power gating opportunity. Using higher number of cut-off transistors and SRAMs, e.g., using two cut-off SRAMs provides higher power gating opportunity since in the case of a partly used SB, it is still able to turn off the power of half of the SB (i.e., three SRAM cells). However, experimental results reveal that the power overhead of using two or more cut-off SRAM compensates the achieved additional opportunity. An average of 27%, 75%, and 4% power saving in LUTs, CBs, and SBs is reported, respectively. It should be noted that the assumed architecture in this study is different from commercial ones. More importantly, medium-oxide transistors are currently used for SRAM cells which significantly reduce their power consumption. State-of-the-art FPGAs leverage multiplexer-based SBs with large output buffers wherein the power of buffers and transistors of the multiplexer structures overshadow the power of SRAMs cells.

In Ref. [50], Hoo et al. suggest a coarse-grained power gating SB architecture with unidirectional segments to cope with static power consumption. The SB multiplexers and associated output buffers are grouped in power gating regions based on their direction. Furthermore, the authors modify the routing algorithm to improve the effectiveness of the proposed architecture. The main objective of the proposed routing algorithm is to increase the unused regions for the purpose of power gating without considerable effect on critical path delay. For this purpose, the base line cost function of VPR router which is based on nodes delay and congestion is modified to reflect the impact of having more power gating regions. Accordingly, the direct relation between the new cost function and the size of available power gating regions hinders the use of large available power gating regions to route the nets, and indirect relationship of new cost function and the number of nets routing through power gating region increases the number of unused power gating region that can be turned off.

Lin et al. reduce the static power consumption of routing SBs via duplicating the routing tracks considering a buffer-based routing architecture [51]. It is noteworthy that state-of-the-art commercial devices use multiplexer-based routing architecture because of its area and performance efficiency. In a fine-grained manner, the associated buffers routing tracks are supplied with either a high or low VDD and the power state (on or off) of each buffer is controlled individually. That is, the configuration cell of each buffer not only determines its utilization state but also controls its connectivity to supply voltage, as well. The authors have also developed new design flow to use the proposed architecture in which the suitable supply voltage is assigned to each routing tree based on the path delay and power. While this architecture achieved power reduction up to 45%, an area overhead of 57%–186% is incurred.

Despite the potential efficiency of power gating technique to reduce the static power consumption, inefficient implementation of power gating can have reverse outcome. In this regard, Seifoori et al. have analyzed the utilization rate of routing resources and found the best power gating granularity based on different SM topologies (i.e., Subset, Wilton, and unidirectional) and different SBs structures including bidirectional and unidirectional ones [52]. For this end, under a detailed investigation of utilization rate of routing resources and different granularities, they found the most efficient power gating architecture (i.e., granularity) as well as the power gating architecture with the best power-area tradeoff. Accordingly, various power gating granularities, including SM level, SB level, and intra-SB level are leveraged. The intra-SB level includes different schemes including one- and two-level power gating. Fig. 20 illustrates one- and two-level power gating for the most fine-grained power gating architecture. Despite higher area overhead of the two-level granularity, it increases the controllability of power gating. The authors have also analytically estimated the power consumption of different power gating architectures based on the utilization rate of resources. The difference between the analytically estimated power consumption and experimentally

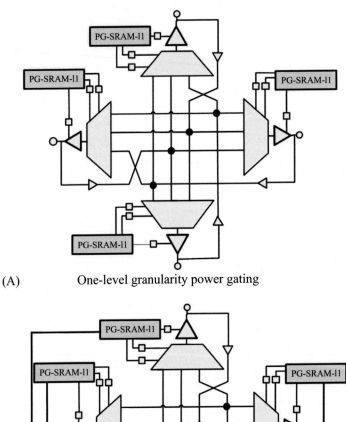

(A)   One-level granularity power gating

(B)   Two-level granularity power gating

Fig. 20 One- and two-level power gating granularity in Ref. [8].

obtained results reveals that the efficiency of power gating architecture depends on other criteria (than solely considering utilization rate) such as SM topology, SB structure, benchmark which affect the distribution of used resources, and also the utilization pattern within each SB.

6. DISCUSSION AND CONCLUSION

Although future horizons of FPGAs and MOSFET semiconductors are getting darker, but researchers and practitioners have proposed various approaches and architectures to shine a bright light at the dark silicon problem in FPGA devices to reduce static power of logic and routing fabrics. However, before escrowing the leadership of computer architecture to FPGA infrastructure, one has to address the major challenges that are lying around the reconfigurable computing in prospect of dark silicon:

1. How to make room for reconfigurable platforms into today's hardware and software codesigns? How FPGA's architecture should evolve to be prevalent in today's huge servers and supercomputers?

2. How to devise and develop appropriate languages, compilers, and CAD tools to support ever-changing and emerging application, workloads, and recently, machine learning algorithms?

3. How to harness FPGA's dark side of FPGA's silicon in order to improve power efficiency, aging, or other objectives?

We believe addressing such important questions is very crucial to advance the FPGA industry and will require deep investigations by researchers in this community in upcoming years.

REFERENCES

[1] I. Kuon, J. Rose, Measuring the gap between FPGAs and ASICs, IEEE Trans. Comput. Aided Des. Integr. Circuits Syst. 26 (2) (2007) 203–215.

[2] N. Hardavellas, M. Ferdman, B. Falsafi, A. Ailamaki, Toward dark silicon in servers, IEEE Micro. 31 (4) (2011) 6–15.

[3] G. Venkatesh, J. Sampson, N. Goulding, S. Garcia, V. Bryksin, J. Lugo-Martinez, S. Swanson, M.B. Taylor, Conservation cores: reducing the energy of mature computations, ACM SIGARCH Comput. Archit. News 38 (1) (2010) 205–218.

[4] C. Chiasson, V. Betz, in: COFFE: fully-automated transistor sizing for FPGAs, IEEE International Conference on Field-Programmable Technology (FPT), 2013.

[5] E. Ahmed, J. Rose, The effect of LUT and cluster size on deep-submicron FPGA performance and density, IEEE Trans. Very Large Scale Integr. VLSI Syst. 12 (3) (2004) 288–298.

[6] C. Maxfield, The Design Warrior's Guide to FPGAs: Devices, Tools and Flows, Elsevier, 2004.

[7] E. Kadric, D. Lakata, A. DeHon, Impact of memory architecture on FPGA energy consumption, in: Proceedings of the 2015 ACM/SIGDA International Symposium on Field-Programmable Gate Arrays, ACM, 2015.

[8] Z. Ebrahimi, B. Khaleghi, H. Asadi, PEAF: a power-efficient architecture for SRAM-based FPGAs using reconfigurable hard logic design in dark silicon era, IEEE Trans. Comput. 66 (6) (2017) 982–995.

[9] H. Zhang, W. Zhaoqiang, W. Yansong, Unit cell model formulation and thermal performance analysis for cross-flow heat exchanger, in: ASME 2016 5th International Conference on Micro/Nanoscale Heat and Mass Transfer, American Society of Mechanical Engineers, 2016.

[10] B. Dally, Power and Programmability: The Challenges of Exascale Computing, DoE Arch-I presentation, 2012.

[11] A.A. Bsoul, J.S. Wilton, H.K. Tsoi, W. Luk, An FPGA architecture and CAD flow supporting dynamically controlled power gating, IEEE Trans. Very Large Scale Integr. VLSI Syst. 24 (1) (2016) 178–191.

[12] S.J. Wilton, A.A. Bsoul, in: An FPGA architecture supporting dynamically controlled power gating, IEEE International Conference on Field-Programmable Technology (FPT), 2010.

[13] N. Hardavellas, The Rise and Fall of Dark Silicon, USENIX, 2012.

[14] A. Kanduri, A.M. Rahmani, P. Liljeberg, A. Hemani, A. Hemani, H. Tenhunen, A Perspective on Dark Silicon, Springer International Publishing, 2017.

[15] W.M. Arden, The international technology roadmap for semiconductors—perspectives and challenges for the next 15 years, Curr. Opin. Solid State Mater. Sci. 6 (5) (2002) 371–377.

[16] S. Borkar, in: The exascale challenge, International Symposium on VLSI Design Automation and Test (VLSI-DAT), 2010, IEEE, 2010.

[17] J. Hruska, The death of CPU scaling: from one core to many—and why we're still stuck, ExtremeTech [online] (2012).

[18] H. Esmaeilzadeh, E. Blem, S.A. Renee, S. Sankaralingam, D. Burger, Dark silicon and the end of multicore scaling, ACM SIGARCH Comput. Archit. News 39 (3) (2011) 365–376.

[19] M.B. Taylor, in: Is dark silicon useful? Harnessing the four horsemen of the coming dark silicon apocalypse, 49th ACM/EDAC/IEEE Design Automation Conference (DAC), 2012, pp. 1131–1136.

[20] J. Shailendra, S. Khare, S. Yada, V. Ambili, P. Salihundam, S. Ramani, S. Muthukumar, in: A 280mV-to-1.2 V wide-operating-range IA-32 processor in 32nm CMOS, Solid-State Circuits Conference Digest of Technical Papers (ISSCC), 2012.

[21] E. Krimer, R. Pawlowski, M. Erez, P. Chiang, Synctium: a near-threshold stream processor for energy-constrained parallel applications, IEEE Comput. Archit. Lett. 9 (1) (2010) 21–24.

[22] V. Govindaraju, C.-H. Ho, K. Sankaralingam, Dynamically specialized datapaths for energy efficient computing, in: 2011 IEEE 17th International Symposium on High Performance Computer Architecture (HPCA), IEEE, 2011, pp. 503–514.

[23] J. Allred, R. Sanghamitra, K. Chakraborty, in: Designing for dark silicon: a methodological perspective on energy efficient systems, Proceedings of the 2012 ACM/IEEE International Symposium on Low Power Electronics and Design (ISLPED), 2012.

[24] J. Cong, M.A. Ghodrat, M. Gill, B. Grigorian, G. Reinman, in: Architecture support for accelerator-rich cmps, Proceedings of the ACM 49th Annual Design Automation Conference (DAC), 2012.

[25] M.J. Lyons, M. Hempstead, G.-Y. Wei, D. Brooks, The accelerator store: a shared memory framework for accelerator-based systems, ACM Trans. Archit. Code Optim 8 (4) (2012) 1–22.

[26] T. Turakhia, R. Bharathwaj, G. Siddharth, D. Marculescu, in: Hades: architectural synthesis for heterogeneous dark silicon chip multi-processors, Proceedings of the 50th Annual Design Automation Conference (DAC), 2013.

[27] M. Spencer, F. Chen, W.C. Cheng, R. Nathanael, H. Fariborzi, A. Gupta, H. Kam, V. Pott, J. Jeon, T.J. Liu, D. Markovic, Demonstration of integrated micro-electro-mechanical relay circuits for VLSI applications, IEEE J. Solid State Circuits 46 (1) (2011) 308–320.

[28] H. Dadgour, K. Banerjee, Design and analysis of hybrid NEMS-CMOS circuits for ultra low-power applications, in: Proceedings of the 44th Annual Design Automation Conference, ACM, 2007, pp. 306–311.

[29] A.M. Ionescu, H. Riel, Tunnel field-effect transistors as energy-efficient electronic switches, Nature 479 (7373) (2011) 329–337.

[30] J. Henkel, L. Bauer, N. Bauer, P. Dutt, S. Nassif, M. Shafique, M. Shafique, N. Wehn, in: Reliable on-chip systems in the nano-era: lessons learnt and future trends, Proceedings of the 50th Annual Design Automation Conference (DAC), 2013.

[31] J. Henkel, L. Bauer, H. Zhang, S. Rehman, M. Hongyan, in: Multi-layer dependability: from microarchitecture to application level, In International Conference on Design Automation Conference (DAC), 2014.

[32] F. Kriebel, S. Rehman, D. Sun, M. Shafique, J. Henkel, in: Aser: adaptive soft error resilience for reliability-heterogeneous processors in the dark silicon era, International Conference on Design Automation Conference (DAC), 2014.

[33] B. Raghunathan, Y. Turakhia, S. Garg, D. Marculescu, in: Cherry-picking: exploiting process variations in dark-silicon homogeneous chip multi-processors, Proceedings of the Conference on Design, Automation and Test in Europe (DATE), 2013.

[34] M. Shafique, S. Garg, J. Henkel, D. Marculescu, in: The EDA challenges in the dark silicon era: temperature, reliability, and variability perspectives, Proceedings of the 51st Annual Design Automation Conference (DAC), 2014.

[35] A. Ahari, B. Khaleghi, Z. Ebrahimi, H. Asadi, B.M. Tahoori, in: Towards dark silicon era in fpgas using complementary hard logic design, International Conference on Field Programmable Logic and Applications (FPL), 2014.

[36] S. Ishihara, M. Hariyama, M. Kameyama, A low-power FPGA based on autonomous fine-grain power gating, IEEE Trans. Very Large Scale Integr. VLSI Syst. 19 (8) (2011) 1394–1406.

[37] R. Ahmed, A.A. Bsoul, S.J. Wilton, P. Hallschmid, R. Klukas, In: High-level synthesis-based design methodology for Dynamic Power-Gated FPGAs, International Conference on Field Programmable Logic and Applications (FPL), 2014.

[38] J.H. Anderson, F.N. Najm, Low-power programmable FPGA routing circuitry, IEEE Trans. Very Large Scale Integr. VLSI Syst. 17 (8) (2009) 1048–1060.

[39] R. Krishnan, J.P. de Gyvez, in: Low energy switch block for FPGAs, 17th International Conference on VLSI Design, 2004.

[40] M. Klein, Power Consumption at 40 and 45 nm, Xilinx, 2009.

[41] J.H. Anderson, F.N. Najm, Active leakage power optimization for FPGAs, IEEE Trans. Comput. Aided Des. Integr. Circuits Syst. 25 (3) (2006) 423–437.

[42] S. Srinivasan, A. Gayasen, N. Vijaykrishnan, in: Leakage control in FPGA routing fabric, Proceedings of the 2005 Asia and South Pacific Design Automation Conference, ACM, 2005.

[43] M. Hasan, A.K. Kureshi, T. Arslan, in: Leakage reduction in FPGA routing multiplexers, 2009 IEEE International Symposium on ISCAS, 2009.

[44] A.A.M. Bsoul, S.J.E. Wilton, in: An FPGA architecture supporting dynamically controlled power gating, 2010 International Conference on Field-Programmable Technology (FPT), IEEE, 2010.

[45] R.P. Bharadwaj, R. Konar, P.T. Balsara, in: Exploiting temporal idleness to reduce leakage power in programmable architectures, Proceedings of the 2005 Asia and South Pacific Design Automation Conference, ACM, 2005.

[46] A. Gayasen, Y. Tsai, N. Vijaykrishnan, M. Kandemir, M.J. Irwin, T. Tuan, in: Reducing leakage energy in FPGAs using region-constrained placement, Proceedings of the 2004 ACM/SIGDA 12th International Symposium on Field Programmable Gate Arrays, ACM, 2004.

[47] C. Li, Y. Dong, T. Watanabe, in: New power-aware placement for region-based FPGA architecture combined with dynamic power gating by PCHM, Proceedings of the 17th IEEE/ACM International Symposium on Low-Power Electronics and Design, IEEE Press, 2011.

[48] A.A.M. Bsoul, S.J.E. Wilton, in: An FPGA with power-gated switch blocks, 2012 International Conference on, Field-Programmable Technology (FPT), IEEE, 2012.

[49] S. Yazdanshenas, H. Asadi, Fine-grained architecture in dark silicon era for SRAM-based reconfigurable devices, IEEE Trans. Circuits Syst. Express Briefs 61 (10) (2014) 798–802.

[50] C.H. Hoo, Y. Ha, A. Kumar, in: A directional coarse-grained power gated FPGA switch box and power gating aware routing algorithm, 23rd International Conference on Field Programmable Logic and Applications (FPL), 2013, IEEE, 2013.

[51] Y. Lin, F. Li, L. He, in: Routing track duplication with fine-grained power-gating for FPGA interconnect power reduction, Design automation conference, 2005. Proceedings of the ASP-DAC 2005. Asia and South Pacific, 2005.

[52] Z. Seifoori, B. Khaleghi, H. Asadi, in: A Power gating switch box architecture in routing network of SRAM-based FPGAs in dark silicon era, IEEE/ACM Design, Automation and Test in Europe Conference (DATE), Lausanne, Switzerland, 2017.

ABOUT THE AUTHORS

Zeinab Seifoori has received her B.Sc. and M.Sc. degrees in computer engineering from Shahed University and SUT, Tehran, Iran, in 2006 and 2010, respectively. From 2015, she has been working toward the Ph.D. degree in computer engineering in *Data Storage, Networks, and Processing* (DSN) Laboratory at the Department of Computer Engineering, Sharif University of Technology. Her research interests include reconfigurable computing and reliability of computer systems.

Zahra Ebrahimi received the B.Sc. degree in computer engineering from *Sharif University of Technology* (SUT), Tehran, Iran, in 2014, wherein she is currently pursuing the M.Sc. degree. She has been with the *Data Storage, Processing, and Networks* (DSN) Laboratory at the Department of Computer Engineering, SUT, as a research assistant for 3 years. Her current research interests include reconfigurable computing and computer-aided design.

Behnam Khaleghi has received his M.S. and B.Sc. degree in computer engineering from Sharif University of Technology (SUT), Tehran, Iran, in 2016 and 2013, respectively. He is currently working as a research assistant in the *Data Storage, Processing, and Networks* (DSN) Laboratory at the Department of Computer Engineering, SUT. He spent the summer 2014 and 2015 as a research assistant at the Chair for Embedded Systems in the Karlsruhe Institute of Technology. His research interests include reconfigurable architectures, computer-aided design, and reliable system design.

Hossein Asadi (M'08, SM'14) received the B.Sc. and M.Sc. degrees in computer engineering from Sharif University of Technology (SUT), Tehran, Iran, in 2000 and 2002, respectively, and the Ph.D. degree in electrical and computer engineering from Northeastern University, Boston, MA, USA, in 2007.

He was with EMC Corporation, Hopkinton, MA, USA, as a Research Scientist and Senior Hardware Engineer, from 2006 to 2009. From 2002 to 2003, he was a member of the Dependable Systems Laboratory, SUT, where he researched hardware verification techniques. From 2001 to 2002, he was a member of the Sharif Rescue Robots Group. He has been with the Department of Computer Engineering, SUT, since 2009, where he is currently a tenured Associate Professor. He is the Founder and Director of the *Data Storage, Networks, and Processing* (DSN) Laboratory, Director of *Sharif High-Performance Computing* (HPC) Center, the Director of Sharif Information and Communications Technology Center (ICTC), and the President of Sharif ICT Innovation Center. He spent 3 months in the summer 2015 as a Visiting Professor at the School of Computer and Communication Sciences at the Ecole Poly-technique Federele de Lausanne (EPFL). He is also the co-founder of HPDS corp., designing and fabricating midrange and high-end data storage systems. He has authored and co-authored more than eighty technical papers in reputed journals and conference proceedings. His current research interests include data storage systems and networks, solid-state drives, operating system support for I/O and memory management, and reconfigurable and dependable computing.

Dr. Asadi was a recipient of the Technical Award for the Best Robot Design from the International RoboCup Rescue Competition, organized by AAAI and RoboCup, a recipient of Best Paper Award at the 15th CSI International Symposium on *Computer Architecture & Digital Systems* (CADS), the Distinguished Lecturer Award from SUT in 2010, the Distinguished Researcher Award and the Distinguished Research Institute Award from SUT in 2016, and the Distinguished Technology Award from SUT in 2017. He is also recipient of Extraordinary Ability in Science visa from US Citizenship and Immigration Services in 2008. He has also served as the publication chair of several national and international conferences including CNDS2013, AISP2013, and CSSE2013 during the past 4 years. Most recently, he has served as a Guest Editor of IEEE Transactions on Computers, an Associate Editor of Microelectronics Reliability, a Program Co-Chair of CADS2015, and the Program Chair of CSI National Computer Conference (CSICC2017).

Printed in the United States
By Bookmasters